New Roads and Street Works Act 1991

MEASURES NECESSARY WHERE APPARATUS IS AFFECTED BY MAJOR WORKS (DIVERSIONARY WORKS)

A Code of Practice

Approved by the Secretaries of State for Transport, Wales and Scotland under sections 84 and 143 of the Act

London: HMSO

June 1992

© Crown Copyright 1992
Applications for reproduction should be made to HMSO
First published 1992
Second impression 1992

ISBN 0 11 551149 0

For an explanation of the asterisk (*) appearing after certain words and expressions in this Code of Practice, the reader should turn to page 2 (Terms and References Applicable in Scotland).

Contents

Executive Overview 1

Terms and References Applicable in Scotland 2

Preface 3

1   Introduction 5

      1.1   Outline of Code 5
      1.2   Applicability of the Code 5
      1.3   Health and Safety 5

2   **Guiding Principles** 7

3   **Background** 9

      3.1   Major Highway Works* 9
      3.2   Major Bridge Works 10
      3.3   Major Transport Works 10
      3.4   Highway* Construction 10
      3.5   Redundant Highways* and Bridges 11
      3.6   Undertakers 11

4   **Planning and Liaison** 13

      4.1   Consultation 13
      4.2   Long-Term Planning 13
      4.3   General Considerations 14
      4.4   Formal Notice 14

5   **Factors Relating to Major Works and Undertakers' Apparatus** 17

      5.1   General 17
      5.2   Change in Depth of Cover 17
      5.3   Change in Lateral Position 18
      5.4   Apparatus at Risk During Construction 21
      5.5   Changes in Type of Highway* Construction 21
      5.6   Protection of Apparatus 22
      5.7   Off-site Works and Network Aspects 22
      5.8   Additional Undertakers' Works 22
      5.9   Highway* Resurfacing 22
      5.10 Overhead and Other Above-Ground Apparatus 23

Figure 1   Change in Depth 25

Figure 2   Change in Lateral Position 27

Figure 3   Apparatus at Risk During Construction 29

Figure 4   Change in Construction 31

6   **Redundant Highways*** 33

      6.1   Formal closure (Stopping Up) 33
      6.2   Change of Use 33

7	**Major Bridge Works and Bridge Replacement Schemes**	**35**
	7.1 General	35
	7.2 Major Bridge Works	35
	7.3 Methods of Undertaking Major Bridge Works	35
	7.4 Methods of Dealing with Undertakers' Apparatus	35
8	**Redundant Bridges**	**37**
9	**General Statement of Financial Agreement**	**39**
	9.1 General	39
	9.2 The Standard Cost Sharing Principle	39
	9.3 Variations on the Standard Principle	39
10	**Conciliation and Arbitration**	**41**
	10.1 Introduction	41
	10.2 Conciliation	41
	10.3 Arbitration	41

Appendix A Types of Undertakers' Apparatus 43

- A1 Gas Industry 43
- A2 Water Industry 44
- A3 Telecommunication and Cable Television Industry 45
- A4 Electricity Supply Industry 47

Appendix B Undertakers' Apparatus - Factors Material to Highway Works* 51

- B1 Gas Apparatus 51
- B2 Water Apparatus 53
- B3 Telecommunication Apparatus 54
- B4 Electricity Cables and Associated Apparatus 59

Appendix C Procedures for Necessary Measures in Relation to Undertakers' Apparatus 65

- C1 Introduction 65
- C2 Preliminary Inquiries 66
- C3 Draft Scheme(s) and Budget Estimates 66
- C4 The Final Detailed Scheme and Detailed Estimates 67
- C5 Formal Notice and Advance Orders 68
- C6 Selection of Contractor and Issue of Main Orders 68
- C7 Construction Stage 68
- C8 Claims 69
- C9 Invoicing, Payment and Financial Monitoring 70

Table 1 - Outline of Procedures in Planning and Implementation of Major Schemes 73

Appendix D Stopping up and Diversion Orders 75

Appendix E Deferment of the Time for Renewal 79

    E1   Deferment of the Time for Renewal 79
    E2   The Bacon and Woodrow Formula 79
    E3   Life of Apparatus 79

Table 2 - Table Applicable to 6 per cent per Annum Interest 81

Appendix F Betterment 85

    F1   Conditions for Allowance 85
    F2   Presentation of Calculations 85

Glossary 87

EXECUTIVE OVERVIEW

1. This Code of Practice addresses major issues identified in the Horne Report associated with circumstances when major highway*, bridge or transport improvement works affect undertakers' apparatus, or when highways* or bridges are to be made redundant.

2. In the case of improvement works, the Code provides information on the needs of the various undertakers and promoting authorities when confronted with the possibility of having to divert or take some other action in regard to undertakers' apparatus. Having provided such information, the Code then gives guidance on the planning of works to ensure that an optimum engineering solution is reached in which the total costs are minimised regardless of who pays.

3. For these objectives to be achieved, there must be close co-operation and co-ordination between the promoting authority and undertaker and the need for this underpins all the provisions of the Code.

4. The Code stresses that diversion only takes place when operational and technical constraints demand it. The Code therefore aims to stimulate thinking by all concerned to explore alternatives to diversion, and where diversion has to take place, to ensure that it is done in a disciplined way.

5. The Code recommends model procedures for dealing with diversionary works from the early planning stages, right through to detailed planning and estimating, execution of works, and final settlement of bills.

6. Regulations under the New Roads and Street Works Act 1991 specify how costs are to be shared. The Code gives more information on the application of these requirements to cope with various diversion scenarios which will be encountered.

7. As a general statement, costs of necessary works incurred will be shared, the promoting authority will bear 82% of the cost of the works and the undertaker 18%. This arrangement gives all parties to diversion a financial incentive to produce a sound solution. The cost sharing agreement is subject to various provisos as detailed in the Regulations and explained in more detail in this Code.

8. In the case of redundant highways* and bridges, the same cost sharing arrangement will apply for moving apparatus existing in the highway* or bridge at the time of formal declaration of redundancy. However, the undertaker will be responsible for the full costs of diverting any additional apparatus placed in the highway* or bridge after the declaration, provided that redundancy actually takes place within the respective five or ten year period.

9. The main body of the Code has been kept deliberately concise to enable major issues to stand out clearly. A series of appendices gives a considerable amount of extra information, explanatory detail and examples of application.

10. Compared to former arrangements, the application of the Code is intended to

    - reduce the number and extent of diversionary works

    - where diversion is nevertheless necessary, to ensure it is done in a way to minimise costs to the community at large

    - to ensure that procedures are adopted at all stages which are readily understood and applied by all concerned

    - to minimise disputes and, if they do occur, provide mechanisms for resolution.

TERMS AND REFERENCES APPLICABLE IN SCOTLAND

References to the road works provisions of the Act are given within the text of the Code of Practice. In the case of terms used which are marked with an asterisk, the reader should note that the following equivalent terms apply in Scotland:

Terms used in the legislation for England and Wales	Equivalent terms used in the Scottish legisation
highway	road
highway authority	roads authority
highway works/works for road purposes	works for road purposes
major highway works	major works for roads purposes
street	road
street authority	road works authority
street works	road works
street works licence	road works permission

The definitions of these and other terms used in the Code of Practice are given in the Glossary.

PREFACE

The following are extracts from the Government's response to Recommendation 67 of the Horne Report on the Review of the Public Utilities Street Works Act 1950:

> "The Government considers that the fundamental objective [where apparatus is affected by highway* works] should be to devise arrangements that will lead to the best overall solution in each case, in particular by minimising the total costs, regardless of who has to meet the costs. It is accepted that the present arrangements may not always achieve this. In particular highway authorities* complain that utilities sometimes require the movement of apparatus when this is not the optimum solution, as utilities try to ensure that they suffer no detriment, but in some cases the movement outweighs a slight detriment from leaving apparatus in its existing position.
>
> The achievement of the optimum solution for each highway* scheme depends upon decisions made by the local staff of both highway authorities* and utilities. The Government considers that the best way of ensuring this must be through those staff working together amicably to obtain the right overall result rather than seeking to protect only the interests of their own organisation. In order to achieve this, a new code of practice will be required stressing this central objective and setting out the considerations that should be taken into account in each case. Such a code of practice could incorporate much of the useful material in the draft Model Guidelines, but it will need to go wider. In particular, all the parties must acknowledge that it will sometimes be right to accept some detriment to their own interests in the overall interest. The code could also include provision for arbitration in the most difficult cases."

These objectives have been put into effect by Parts III and IV of the New Roads and Street Works Act 1991 (hereinafter referred to as 'the Act'). Section 84 (for England and Wales) and section 143 (for Scotland) place a new duty on highway*, bridge and transport authorities and undertakers, where an undertaker's apparatus in a street* is or may be affected by certain major works, to "take such steps as are reasonably required -

(a) to identify any measures needing to be taken in relation to the apparatus in consequence of, or in order to facilitate, the execution of the authority's works,

(b) to settle a specification of the necessary measures and determine by whom they are to be taken, and

(c) to co-ordinate the taking of those measures and the execution of the authority's works,

so as to secure the efficient implementation of the necessary work and the avoidance of unnecessary delay."

The Secretary of State is empowered under sections 84(2) and 143(2) of the Act to issue or approve a code of practice giving practical guidance as to those matters and the steps to be taken by the authority and undertaker. If the authority or undertaker fails to comply with an agreement between them as to any of those matters, either party is liable to compensate the other in respect of any loss or damage resulting from the failure (sections 84(4) and 143(4)).

The Secretary of State is also empowered by sections 85 and 144 of the Act to make Regulations as to how allowable costs for the necessary works will be borne by the parties concerned. The Street Works (Sharing of Costs of Works) Regulations 1992, SI 1992 No 1690, and the Road Works (Sharing of Costs of Works)(Scotland) Regulations 1992, SI 1992 No 1672 (S.159), have been made under these powers and their requirements incorporated in this Code.

This Code of Practice is based on guidance prepared jointly by representatives of highway authorities and the major utilities after the publication of the Government Response[1] to the Horne Report[2].

The Code of Practice has been approved by the Secretaries of State for Transport, Wales and Scotland under their powers in the Act for the purposes of sections 84 and 143; it gives practical guidance about the matters referred to in the previous paragraphs and the steps to be taken by the authority and the undertaker.

The Code was prepared by a working party of the Highway Authorities and Utilities Committee (HAUC), and was the subject of extensive consultation with interested organizations. The following were represented on the working party: the National Joint Utilities Group (comprising British Gas plc, British Telecommunications plc, Mercury Communications Limited, and the water and electricity supply industries in England, Scotland and Wales); the local authority associations (comprising the Association of County Councils, the Association of District Councils, the Association of Metropolitan Authorities, and the Convention of Scottish Local Authorities); British Railways Board; and the Department of Transport.

The Regulations and Code of Practice will come into operation on 1 January 1993.

Enquiries about the Regulations or this Code of Practice should in the first instance be addressed to the Department of Transport, NGAM2, Room 3.14, 2 Monck Street, London SW1P 2BQ. It should be understood that any response that may be given to such enquiries is without prejudice to the rights of parties in a dispute to arbitration under the Act or the Regulations: in this connexion, the reader is also referred to section 10, sub-section 10.3 of this Code.

[1] Department of Transport
Public Utilities Street Works
The Government Response to the Horne Report on the
Review of the Public Utilities Street Works Act 1950 -
July 1986
[ISBN 0 11 550780 9]

[2] Department of Transport
Roads and the Utilities
Review of the Public Utilities Street Works Act 1950
Report to the Secretary of State for Transport, the Rt Hon. Nicholas Ridley MP, of a
Committee chaired by Professor Michael R Horne OBE - November 1985
[ISBN 0 11 550729 9]

1 INTRODUCTION

1.1 Outline of Code

This Code of Practice deals with the question of what action, if any, is required where major highway works*, major bridge works or major transport works affect undertakers' apparatus in or above streets.

The Code has been designed to provide background information which will enable both the authority and undertaker to have an understanding of each other's roles and responsibilities, and provide technical information to enable a sensible appreciation of the issues to be gained. Guidance is provided so that mutual agreement can be reached in individual circumstances but this guidance must be seen against the desire to avoid diversionary works wherever possible. Such guidance cannot be translated into advice which would apply, for example, to the laying of new apparatus within a street*.

Section 2 sets out the four basic guiding principles against which all decisions should be judged.

Section 3 - Background, provides a simplified description of the role and responsibilities of various authorities and undertakers. Appendix A sets out a technical description of a variety of undertakers' apparatus.

Section 4 - Planning and Liaison, describes procedures to be adopted with a view to minimise the need for diversionary works in the future.

Sections 5 to 8 give practical advice on the need or otherwise to divert apparatus due to Promoter's works within a street* or bridge. Advice is also given with regard to redundant highways* and bridges. The Code includes procedures for the efficient implementation of measures needed to be taken in relation to apparatus and the avoidance of unnecessary delay.

Section 9 deals with the general financial arrangements, and Section 10 sets out a procedure for conciliation and arbitration in the event of any unresolved disagreement.

More detailed information is included in appendices which describe the nature of undertakers' apparatus likely to be found in streets*, and the constraints that such apparatus place on actions unaffected by major highway*, bridge and transport works. More detailed information on the application of the financial arrangements is also given in the appendices.

1.2 Applicability of the Code

The Code has a slightly different range of application compared to other aspects of the Act in respect of street works* in the highway*.

　　i.　　It only applies to undertaker's apparatus laid in a maintainable highway* by virtue of a statutory right.

　　ii.　　The Code applies to pipes, ducts, cables, and other apparatus in the highway*, as well as to overhead lines and other apparatus on or above ground, provided they are within the boundary of the highway*. For example, in the case of a cross-country pipeline or overhead line, the Code provisions will only apply to any section which is within the boundary of the highway*.

　　iii.　　The Code also applies to any necessary off-site works.

1.3 Health and Safety

Although the Code provides practical guidance on health and safety issues, it is not intended as an authoritative interpretation of health and safety legislation.

2　GUIDING PRINCIPLES

Four major points of principle which have been embodied into this Code of Practice -

i. There is to be a general presumption against moving apparatus, other than as set out in the Code.

ii. Total costs should be minimised consistent with good practice, regardless of who has to meet the costs.

iii. Staff must work together to obtain the optimum solution rather than seeking to protect only the interest of their own organisation.

iv. All parties must acknowledge that it will sometimes be right to accept some detriment to their own interests in the overall interest.

3 BACKGROUND

This section provides a simplified explanation of the work undertaken by highway*, transport, and bridge authorities and undertakers - it is **not** intended to be exhaustive or prescriptive.

3.1 Major Highway Works*

The role of the highway authority* to maintain, reconstruct and improve the highway* system in the interests of the public must be recognised and every endeavour should be made to keep overall costs of such works to a minimum. In general, the law allows the public to pass and repass along the highway* and to have continued access to land and properties adjoining the highway* except as may be limited by specific legislation. Such rights must be protected.

For England and Wales 'Major highway works' within the meaning of section 86(3) of the Act are:

"works of any of the following descriptions executed by the highway authority in relation to a highway which consists of or includes a carriageway -

(a) reconstruction or widening of the highway,

(b) works carried out in exercise of the powers conferred by section 64 of the Highways Act 1980 (dual carriageways and roundabouts),

(c) substantial alteration of the level of the highway,

(d) provision, alteration of the position or width, or substantial alteration in the level of a carriageway, footway or cycle track in the highway,

(e) the construction or removal of a road hump within the meaning of section 90F of the Highways Act 1980,

(f) works carried out in exercise of the powers conferred by section 184 of the Highways Act 1980 (vehicle crossings over footways and verges),

(g) provision of a cattle-grid in the highway or works ancillary thereto, or

(h) tunnelling or boring under the highway."

For Scotland 'Major works for road purposes' within the meaning of section 145(3) of the Act are:

(j) reconstruction or widening of the road,

(k) substantial alteration of the level of the road,

(l) provision, alteration of the position or width, or substantial alteration in the level of a carriageway, footway or cycle track in the road,

(m) the construction or removal of a road hump within the meaning of Section 40 of the Roads (Scotland) Act 1984,

(n) works carried out in exercise of the powers conferred by Section 63 of the Roads (Scotland) Act 1984 (new access over verges and footways),

(p) provision of a cattle-grid in the road or works ancillary thereto, or

(q) tunnelling or boring under the road.

As the Code applies to 'highways* which consist of or include a carriageway' it does not apply to footpaths or bridleways which are not associated with a carriageway.

Major highway works* as defined are undertaken either by a private contractor under contract to the highway authority*, by the highway authority* itself through its Direct Labour Organisation (DLO) or other body under licence from the highway authority. Financial claims may be made by contractors (including DLOs) against their employer if they suffer delays to their programmes due to factors outside their control which would include the failure of an undertaker to meet an agreed programme.

Some works in the highway* may be undertaken in whole or part by a private developer. Insofar as this Code is concerned, the main impact is on financial arrangements as described in section 9.

3.2 Major Bridge Works

'Major bridge works' within the meaning of the Act in section 88(2) for England and Wales and section 147(2) for Scotland are works for the replacement, reconstruction or substantial alteration of a bridge.

Bridge works may be undertaken as an adjunct to highway works*, railway or other works. There may also be a need to reconstruct, alter or strengthen bridges, to meet current loading or other requirements. Such changes will almost certainly affect any undertakers' apparatus laid in the bridge.

For the purpose of the Code, a bridge means a structure carrying a highway* or a structure over a highway*. The bridge may be the responsibility of any one of a variety of authorities, e.g. the highway authority*, British Railways Board, British Waterways, London Underground Limited. Bridge structures may be of arch, girder or deck type, the materials may be masonry, brick, steel or concrete (reinforced or prestressed).

In the case of a bridge carrying the highway*, undertakers have certain powers to lay apparatus in the highway* and the presence of this apparatus will affect any work undertaken to the bridge structure. By virtue of section 88 and section 147 of the Act, any statutory right to place apparatus in the street includes the right to place apparatus in, and attach apparatus to, the structure of the bridge. However, section 88(5)(b) (for England and Wales) and section 147(5)(b) (for Scotland) require the undertaker to comply with any reasonable conditions the bridge authority may wish to impose for the protection of the bridge or access to it. Some undertakers' apparatus may be located elsewhere on the bridge structure and subject of a different form of agreement. In the case of a bridge **over** the highway*, apparatus laid in the highway* may be affected by work undertaken to the bridge foundations, piers or abutments.

3.3 Major Transport Works

Organisations which have statutory authority to carry on a railway, tramway, dock, harbour, pier, canal or inland navigation undertaking - collectively known as 'transport undertakings' in the Act in section 91(1)(b) (for England and Wales) and section 150(1)(b) (for Scotland) - may also undertake works in their property which affect undertakers' apparatus in streets*. Where such works are 'major transport works' as defined in section 91(2) and section 150(2) as "substantial works required for the purposes of the transport undertaking and executed in property held or used for the purposes of the transport undertaking", the transport authority and undertaker are required to identify and agree necessary measures to be taken in relation to the undertaker's apparatus in the street*. The procedures in this Code should be followed in the same way in such cases; for example, an important example of 'major transport works' affecting undertakers' apparatus will be bridge works and so will be dealt with in the same way as 'major bridge works' in the preceding section 3.2.

3.4 Highway* Construction

The structural design of highways* seeks to satisfy three main objectives:

   i.   To ensure that the load-carrying capacity of the subgrade is not exceeded.

   ii.   To provide a waterproof surface to protect the subgrade from damage by water penetration.

   iii. To provide a satisfactory wearing surface for pedestrian and vehicular traffic.

Modern highway* design is usually based on nationally-accepted recommendations which involve appropriate tests on the subgrade and an assessment of the volume of traffic having standard axle weights likely to use the road. Special design requirements may apply where abnormal loads are anticipated.

With proper structural design of the highway*, no unacceptable loading should be transmitted to undertakers' apparatus beneath the subgrade.

3.5 Redundant Highways* and Bridges

Highways may become redundant because of redevelopment or the construction of an alternative or diverted road (see section 6 of this Code). Similarly, bridges may become redundant because the highway* has closed or because, for example, a railway or canal above or below the highway* has closed. A highway* or bridge becoming redundant will have implications for an undertaker with apparatus in it (see section 8 of this Code).

3.6 Undertakers

Undertakers have a fundamental requirement, which may be imposed by statute, to maintain adequate service while their apparatus is modified or diverted. Any diversion or protective works will have technical features peculiar to the service affected. These requirements may, according to circumstances, be satisfied by a temporary interruption to the service, by diversion to a temporary route, or by permanent re-routing of the service.

Existing apparatus in the highway* may include -

    i.   Underground apparatus comprising cables, pipes, ducts and associated components.

    ii.   Buried apparatus with a surface access such as chambers, manholes and valves. As the networks of pipes and cable generally follow the road layout, junction boxes, chambers and manholes are often sited at, or near to, road intersections.

    iii.   Apparatus above ground such as poles, pylons, overhead cables, interconnection cabinets, section pillars, pressure regulators, etc.

    iv.   Special structures such as pipe or cable bridges and tunnels.

Undertakers' apparatus within the highway* is likely to perform one of two principal functions - either service mains/cables providing a direct service to properties or trunk mains/cables which form part of the main distribution system. This distinction is important as it can influence decisions as to the action to be taken if they are affected by highway works*.

More information on the types of undertakers' apparatus is set out in Appendix A.

4 PLANNING AND LIAISON

4.1 Consultation

When major highway*, bridge or transport works involve alteration to undertakers' apparatus, effective planning and liaison between the authority and undertaker is of great importance.

The designer of a scheme needs to recognise the importance of early consultation with undertakers before planning progresses to the detail stage. The presence of undertakers' apparatus can have profound effects on the practicality and economics of many schemes. It is possible, for example, that minor alterations to the intended layout may lead to considerable savings in the cost of diverting undertakers' apparatus by minimising the amount of necessary work, or perhaps in some cases, avoiding the need to alter the undertakers' apparatus. It must also be recognised that most schemes affect the apparatus of more than one undertaker and optimising the final highway layout to reduce the overall cost may be a complex task.

Detailed procedures for the various stages of the planning and implementation of major highway works* affecting undertakers' apparatus are given in Appendix C.

Undertakers will also need to establish liaison with the authorities when proposing to lay or modify apparatus in the highway* to minimise problems during the work or at a subsequent time.

4.2 Long-Term Planning

Considerable benefits could accrue if all parties share long-term plans (provided commercial sensitivity is not affected). This section suggests some sources and uses of such information.

4.2.1 Highway* And Other Relevant Authorities

Information sources include:

  (a) Most English highway authorities, and Scottish local roads authorities produce an annual Transport Policy and Programme (TPP) which will normally identify:

   i. Major highway works* on local roads programmed up to 10 years ahead.

   ii. Other highway works* on local roads programmed for 2 to 5 years ahead.

   A copy of the relevant parts of the TPP should be sent to each local undertaker. Whether or not a proposed scheme will proceed will depend on financial approval.

  (b) In Wales, details of schemes over £3m are published by the Secretary of State for Wales. Details of schemes under £3m in Wales are published by agent authorities or highway authorities*.

  (c) Details of major trunk road programme schemes in preparation are set out in reports prepared from time to time by the Department of Transport and The Scottish Office and published by and obtainable from the HMSO.

  (d) Other bridge and transport authorities should send a copy of their long-term plans for major works which might affect undertakers' apparatus to each local undertaker where appropriate.

4.2.2 Undertakers

Undertakers should consult the highway* and other relevant authorities on their longer-term planning intentions and particularly when it is proposed to place apparatus which would be expensive to divert should it be affected by major works in the future. Regard should also be taken of the existence of any protected streets*. These are streets* designated under section 61 of the Act (for England and Wales) and section 120 of the Act (for Scotland), in which the consent of the street authority* is required for the placing of new apparatus by undertakers.

In addition to planning and liaison on specific major works, the location of new apparatus, or relocation of existing apparatus should, as far as is possible, make provision for future improvements to the highway*. Such improvements are most likely to be needed along the more heavily trafficked roads, e.g. the classified roads together with those unclassified roads which form district or local distributors. Apparatus placed within these roads should wherever possible be located so as to minimise the need for future alteration. For example, it is desirable that apparatus in close proximity to road junctions where later changes to turning radii are possible should be laid at carriageway depth, as set out in Appendix B. Equally, frames and covers, surface apparatus, etc. should be located away from the minor road kerbline. Such practices are likely to afford the highway authority* some opportunity of improving turning radii at junctions without the consequential need to seek relocation of undertakers' apparatus.

4.3 General Considerations

At the early stage of a scheme being studied, the authority is unlikely to have sufficient information for detailed plans to be produced. However, by submitting preliminary drawings of their proposals they can obtain copies of the record plans from undertakers which will assist in planning.

It is sometimes necessary for undertakers to carry out work remote from the highway*, bridge or transport scheme in order that diversionary works can proceed. Such works may enable the undertaker's service to be re-routed and the apparatus in the way of the major works to be disconnected. In many situations, it will be necessary to undertake some or all of the undertaker's alterations prior to the commencement of the main works contract.

An important aspect for undertakers is that they must be able to programme any interruption of service to their customers and close co-ordination between all parties involved will be necessary to ensure this.

The designers of major highway*, bridge or transport works will need to take account of aspects of undertakers' work which may require to be progressed well in advance of contracts being placed. For example, some items of undertakers' apparatus may not be readily available. The acquisition of land or the negotiation of any necessary wayleaves or easements may also be a protracted procedure. Where the undertaker has to have special facilities constructed, such as sub-stations, it will be necessary to have access to the sites at the earliest possible stage.

4.4 Formal Notice

4.4.1 The General Case

It is the intention of this Code that authorities and undertakers should plan their work jointly so as to avoid conflict and unnecessary diversionary works. However, it will be necessary to retain a system of formal notification.

Before major highway*, bridge or transport works can proceed, the authority will be required to issue a formal notice to undertakers under section 85 of the Act (for England and Wales) or section 144 of the Act (for Scotland), and the highway authority* to register the information on their street works* register under section 53 of the Act (for England and Wales) or section 112 (for Scotland). Undertakers should acknowledge receipt of this notice. Appendix C of this Code gives detailed procedures for planning and liaison in association with these works.

In addition, an undertaker is required to give advance notice to relevant authorities of his intention to carry out more major street works* under section 54 of the Act (for England and Wales) or section 113 of the Act (for Scotland). An authority may therefore object by serving a notice on the undertaker under section 85 or section 144 of the Act, stating their intention to execute major highway*, bridge or transport works.

When serving such a 'counter notice', the authority should demonstrate their reasonable expectation of undertaking the works within five years of the issue of their notice. A counter notice must be received not more than one month from notice being given by the undertaker.

Every effort should be made to reconcile the undertaker's and the authority's positions, but notwithstanding the serving of such a counter notice, the undertaker may nevertheless choose to proceed with its major works. However, where it does so in the case of the laying of additional apparatus and the major highway*, bridge or transport works are commenced within a five-year period, then the undertaker will be responsible for the cost of any necessary diversionary works or protection of such additional apparatus.

4.4.2 The Special Case of Redundant Bridges

In the case of a redundant bridge, the relevant authority should normally issue a formal declaration of redundancy when demolition is expected to occur within ten years of such a declaration. On receipt of a formal declaration, the undertaker must respond with a statement of the apparatus in the bridge. The absence of a declaration of redundancy does not prevent the authority from issuing a 'counter notice' in response to an undertaker.

When issuing either a declaration of redundancy or a 'counter notice', the authority would be expected to demonstrate the reasonableness of their expectation to undertake demolition within ten years e.g. for local road bridges, evidence by way of reference in a TPP.

Again, every effort should be made to reconcile the undertaker's and authority's positions but the undertaker may nevertheless choose to proceed with his works. However, if the undertaker proceeds with the laying of additional apparatus in a redundant bridge, and demolition takes place within the ten year period, then the undertaker will be responsible for the full cost of diversionary works of such additional apparatus.

5 FACTORS RELATING TO MAJOR WORKS AND UNDERTAKERS' APPARATUS

5.1 General

Undertakers' apparatus affected by major highway*, bridge or transport works may be -

  i. left in situ and unmodified;

  ii. left in situ but protected or strengthened;

  iii. temporarily diverted or protected during the major works;

  iv. moved to a new position; or

  v. abandoned and new apparatus provided in a new position.

Items (ii) to (v) are the most common options in what are called 'necessary measures' in the context of sections 84 and 85 of the Act (for England and Wales), or sections 143 and 144 of the Act (for Scotland).

Every effort should be made to leave the plant in situ and protected where necessary (see also section 5.6 below).

There will always be clear-cut cases where undertakers will have no option but to divert their apparatus because, unless this is done, the works cannot go ahead. In other cases there are a number of key factors to be considered, either individually or collectively, when deciding whether or not to divert. These factors are as follows -

  i. change in depth of cover

  ii. change in lateral position

  iii. apparatus at risk during construction

  iv. change in type of highway* construction.

These four factors are given in more detail in sections 5.2 to 5.5 and are shown diagrammatically in Figures 1 to 4 (see pages 25 to 31).

It will also be necessary to consider the wider issues of the undertaker's network as described in section 5.7.

In deciding how best to deal with undertakers' apparatus affected by major highway*, bridge or transport works, undertakers and authorities need to take full account of health and safety requirements.

5.2 Change in Depth of Cover

(See Figure 1, page 25)

In some cases the proposed level of the road is such that the apparatus has to be moved to allow excavation in the new level. In other cases, the need for diversion is almost as clear cut because the apparatus, if left, would be very shallow or very deep, neither of which is acceptable.

5.2.1 Highway* Engineering Considerations

It is rarely acceptable to have the apparatus located within the footway or carriageway construction, and preferably not within 75 mm of it except in special circumstances. The highway authority* may well require a greater depth of cover but, in so doing, may incur a greater degree of diversionary work.

5.2.2 Apparatus Considerations

Where the resulting depth of cover is outside the standard values defined for each undertaker in Appendix B, there are a number of general factors to be considered when deciding whether diversion is necessary or whether protection is possible. These factors include -

    i.   material of the apparatus

    ii.   type of construction of the apparatus e.g. welded, jointed, etc.

    iii.   the structural condition of the apparatus

    iv.   the duty or rating of the apparatus

    v.   external mechanical loading on apparatus

    vi.   length of apparatus affected

    vii.   ease of access

    viii.   the effect on services to customers and on ancillary plant

    ix.   the type of highway* construction, structural stability and traffic volume

    x.   the effects of climate, i.e. freezing and ground movement

See Appendix B for further detail.

5.3 Change in Lateral Position

(See Figure 2, page 27)

There are two situations where, as a result of highway works*, apparatus may have to be considered for diversion because of a change in lateral position -

    i.   apparatus already under the carriageway which moves to a different position within the carriageway; and

    ii.   apparatus not under the carriageway which would become situated under the carriageway.

The decision as to whether the apparatus should be moved depends on a number of factors that are considered in detail in the following sub-section -

    i.   highway* engineering aspects (section 5.3.1)

    ii.   apparatus criteria (section 5.3.2)

    iii.   traffic disruption problems (section 5.3.3)

    iv.   frequency of access to apparatus (section 5.3.4)

    v.   safety of undertakers' operatives (section 5.3.5)

A general appraisal of these factors is covered below and aspects relevant to each undertaker are considered in Appendix B.

It is important to ensure that each of these factors as illustrated in Figure 2 is appraised before deciding whether or not to divert. Generalising, there are a number of problems that could arise with highway works* causing a change in lateral position of apparatus.

Apparatus is normally laid at a shallower depth in the footway or verge than in the carriageway. If the widened carriageway encompasses the former footway or verge this could cause problems of the adequacy of cover (see section 5.2 above for change in depth criteria). Water apparatus, because of its potential for freezing, is normally laid at the same depth regardless of location. Because of the likely frequency of access, apparatus with service connections or ancillary apparatus (such as electricity link boxes) is normally situated in the footway or verge. If the widened carriageway encroaches on the footway or verge, operational problems could arise.

Problems can arise when apparatus is so positioned in the highway* that the traffic flow is seriously inhibited when access to the apparatus is required. This is a particular problem if apparatus is situated in the centre of a two-lane road.

5.3.1 Highway* Engineering Considerations

Changes in lateral position should not normally result in apparatus being -

   i. located under the line of the kerb,

   ii. located adjacent to the kerb line so as to interfere with the installation of gullies, drainage connections, traffic signs and street lighting columns, or

   iii. located at the back of the footway if there is a requirement to erect lighting columns in this position.

Regardless of apparatus considerations, a cover of not less than 75 mm below the footway or carriageway formation level should be provided to give adequate clearance for highway* operations.

5.3.2 Apparatus Criteria

The major characteristics of the apparatus of each undertaker in so far as these influence decisions on diversion of apparatus are given in Appendix B. There are however a number of general features which are included in this section -

   i. The distinction between 'service' mains/cables providing a direct service to properties and 'trunk' mains/cables which form part of the main distribution system. The terminology for 'service' and 'trunk' varies between undertakers and is discussed further in Appendix B. One definition of 'service pipe' or 'service line' is given in paragraph 7 of Schedule 4 to the Act (for England and Wales) and paragraph 7 of Schedule 6 to the Act (for Scotland) (in the context of submission of plans and sections for works on streets* with special engineering difficulties).

   In general, access is likely to be required much more frequently to the 'service' element of the network compared to the 'trunk' element.

   ii. Apart from the depth of cover question discussed in section 5.2 above the change of location from footway/verge to carriageway has other significant implications. For example, many of the access points - jointing chambers, valve housings, link boxes etc. are designed specifically for footway use and could not be permitted to be used in the carriageway without alteration.

   iii. The decision on diversion of apparatus will be influenced, on the one hand, by the apparatus itself, the materials used in the construction, its age, condition, and susceptibility to disturbance. On the other hand, the characteristics of the loading which would be applied as the result of a possible change of situation are also important both in magnitude and frequency of occurrence.

   iv. The consequences of a pipe or cable failure may be such that access must be possible without delay. In the case of an electricity cable, a fault may lead to high earth currents, a gas leak carries with it the risk of fire or explosion, a water leak may lead to flooding or undermining of plant or buildings, and the loss of vital telecommunications circuits will require urgent attention.

   v. As a general statement, the apparatus of the piped undertakers is much more susceptible than the apparatus of the cable undertakers to disturbance in the form of vibration, excess loading, subsidence or other ground movement.

vi. It may be possible to arrange with the undertaker to slew, raise or lower cable to some extent. Such movement is not normally possible with pipes even if plastic, and in the case of cables, in the vicinity of joints.

vii. Continuity of supply. In most cases, it will be necessary for undertakers to maintain continuity of supply and many techniques have been developed to achieve this aim. In the case of diversions, both on-site and off-site works may be necessary.

5.3.3 Traffic Disruption and Road Safety

The highway authority* has a responsibility to maintain the safe and free flow of traffic and pedestrians along the highway* and will therefore be concerned about the location of service boxes, chambers and other apparatus.

Depending upon both the frequency of access and length of time involved, the highway authority* should evaluate the costs and benefits of the various options in terms of traffic disruption and road safety arising from the location of apparatus. There will be more concern about traffic disruption on main roads than minor roads and account will also need to be taken of the capacity of the road and the availability of alternative traffic routes.

The existence of traffic-sensitive streets* will also be relevant to the view formed by the authority and undertakers; additional disruption on such streets* would be less acceptable than on other routes. Road safety concern will be pertinent, particularly if it proves difficult to apply acceptable temporary traffic signing and lane closures.

Whilst traffic disruption may not be an overriding factor warranting the diversion of services, the authority should be prepared to discuss with the undertakers related factors such as frequency of access and safety of operatives with a view to reaching a joint conclusion.

5.3.4 Frequency of Access

Undertakers will need to evaluate the cost and operational implications of access to existing apparatus in a new situation.

The case for resiting apparatus is more cogent when it is likely to require frequent access for installation, inspection, maintenance and/or repair than when access is only required infrequently. Frequent access is more likely for service mains providing a direct service to properties. The case for resiting may be less where access problems to existing apparatus have been accepted. The likely frequency of access will depend upon the precise nature of the apparatus concerned. This subject is dealt with in more detail in the Appendices A and B.

For some undertakers, the provision of side-shaft entries to manholes may be possible.

The general indications in favour of moving apparatus out of the carriageway are strongest when the road is an urban through route or a rural major road and access to the apparatus is frequently required. They are less strong in the case of other roads and when access is not frequently required, when it may often prove acceptable and in the general public's interest to avoid the expense of removal.

5.3.5 Safety of Staff

The following factors should be taken into account when considering the working conditions for staff -

i. The likely density and speed of traffic on the road and the visibility of the works site.

ii. Frequency and duration of periods of access.

iii. The size of working party who are most likely to be required to gain access to the apparatus and the signing, lighting and guarding equipment which they are likely to carry.

iv. It must be practical to install appropriate signing, lighting and guarding at the site, and the undertaker must secure that the works are adequately signed, lit and guarded in accordance with section 65 of the Act (for England and Wales) or section 124 of the Act (for Scotland), and comply with any directions given by the traffic authority.

v. The ultimate responsibility for staff safety rests with the undertaker who has also to satisfy the requirements of the Health and Safety at Work etc Act 1974. In exercising this responsibility, the undertaker should take into account the views of both the police and the highway authority*.

5.4 Apparatus at Risk During Construction

(See Figure 3, page 29)

Construction factors to be considered in deciding whether apparatus is at risk include -

i. removal of the footway or carriageway construction;

ii. construction plant crossing or working in the vicinity of the apparatus;

iii. the undermining or removal of side support of the apparatus;

iv. deep excavations adjacent to the apparatus; and

v. piling or ground consolidation operations nearby.

It may be possible to take steps on site to avoid or minimise risk, for example, by providing plant crossing points. Similarly some types of apparatus can be undermined for short sections and small lateral movements of cables and ducts are sometimes possible. In other cases, undermining is not acceptable and lateral movement is not possible, as for example with cast iron pipes. Each case will have to be carefully reviewed and the conclusion reached will depend upon the nature of the apparatus and the effects of the construction phase.

The apparatus may be protected or temporarily supported in situ, or temporarily moved to a new location during the construction stage or else permanently moved to a new position. In deciding which option to pursue it should be checked whether protective measures or temporary diversions would be more expensive than the alternative of permanent re-siting. In certain circumstances, the risk of damage may be such that equipment has to be re-sited to safeguard continuity of service during the construction stage. The potential effect of the construction works on ancillary apparatus has also to be determined.

Undertakers and contractors must be reminded of their responsibility to safeguard apparatus during construction works and avoid any undue risks. Full account must be taken of the recommendations of Health and Safety Guidance Note HS(G)47 'Avoiding danger from underground services'.

5.5 Changes in Type of Highway* Construction

(See Figure 4, page 31)

A change in the type of highway* construction could have implications for the need to divert apparatus in three circumstances -

i. if the clearance of the apparatus is reduced to less than 75 mm below the level of the footway or carriageway formation level (see section 5.2 above);

ii. if a flexible road construction is replaced by concrete; or

iii. if expensive paving/surfacing is used in, for example, a pedestrianised area.

The highway authority* is unlikely to require or pay for apparatus to be diverted in either circumstance (ii) or (iii), but an undertaker may consider the apparatus should be diverted because of the increased difficulty of maintaining and inspecting apparatus and reinstating excavations.

5.6 Protection of Apparatus

Where, as the result of highway works*, a pipe, cable, or duct is at less than the minimum depth recommended for that particular apparatus it may be possible, for certain types of apparatus, to leave it in situ but provide it with additional protection. In particular, these circumstances may arise over relatively short lengths such as, for example, with the construction of laybys. Protection will normally be in the form of a concrete raft placed over the apparatus. This raft may have to be separated from the apparatus by a flexible cushion. Jointing chambers and manholes may also require modification if apparatus is left at less than minimum depth. Where carriageways are of rigid construction, some form of flexible cushion should be left between the underside of the road construction and any protective raft. It is important to recognise however that the provision of protective measures is sometimes more expensive than diversions and can lead to later access problems and increased maintenance costs.

The specification and implementation of protective measures should be agreed between the undertaker and highway authority* as will any future maintenance liabilities.

5.7 Off-site Works and Network Aspects

All undertakers' apparatus forms part of a network and when part of that network is affected by highway works*, action taken by the undertaker will have to take into consideration the network aspect. This section draws attention to some possible consequences of this -

i. Any new length of pipe or cable will have to be compatible with that section of the network. In some cases different sizes and materials may have to be used and in other cases apparatus which is not currently manufactured may have to be specially obtained which will adversely affect costs and timescales.

ii. In other situations, the apparatus used for the diverted section may not be identical to that which had to be abandoned. Smaller capacity pipes or cables may be installed if current information on likely developments shows that some of the existing capacity is unlikely to be required in the future. The converse may also apply, in which case the undertaker would take the opportunity of highway works* to increase at his expense the capacity of the apparatus.

iii. The network aspect also means that replacement of short lengths of pipe or cable does not give quantifiable benefits to the undertaker as the capacity of the network will be determined by the lowest capacity section.

iv. In many instances, it will be necessary for the undertaker to execute off-site works before, during or after highway works*. The costs incurred in such off-site work may exceed those of on-site works.

v. In limited circumstances it may be possible for the undertaker to re-arrange the network such that all or part of the section affected by the highway works* could be abandoned completely. In such a case, all costs would relate to off-site works to effect the reconfiguration of the network.

5.8 Additional Undertakers' Works

The undertaker will wish to consider the opportunity presented by highway works* to modify or extend his network at his own expense.

Even where apparatus is not directly affected by the highway works*, consideration should be given to undertaking renewal or relocation, particularly if the existing apparatus is at sub-standard depth.

Where the undertaker decides to carry out additional work, early consultation with the authority will be necessary to enable the works to be programmed with the highway works*.

5.9 Highway* Resurfacing

Highway works* which involve relevelling of undertakers' frames and covers should normally be undertaken by the highway authority* to an agreed specification which must incorporate the surface profile requirement in S2.2

of the 'Specification for the Reinstatement of Openings in Highways' code of practice. When requested by the undertaker, the highway authority* must ensure that unauthorised access cannot be gained to apparatus, for example, by the provision of suitable temporary covers.

Any special materials required for the purpose will be supplied (and fitted if necessary) free-of-charge by the undertaker, as will new frames and covers if the existing ones are worn or damaged.

These works will not be treated as diversionary works in the context of sharing of costs in section 9. Resurfacing should not normally justify diversion just because of a marginal increase in depth of cover.

5.10 Overhead and Other Above-Ground Apparatus

Sections 5.1 to 5.9 are written in the context of underground apparatus; the principles however are applicable where appropriate to situations where overhead lines, pipe and cable bridges and other above-ground apparatus are affected by highway*, bridge or transport authority works.

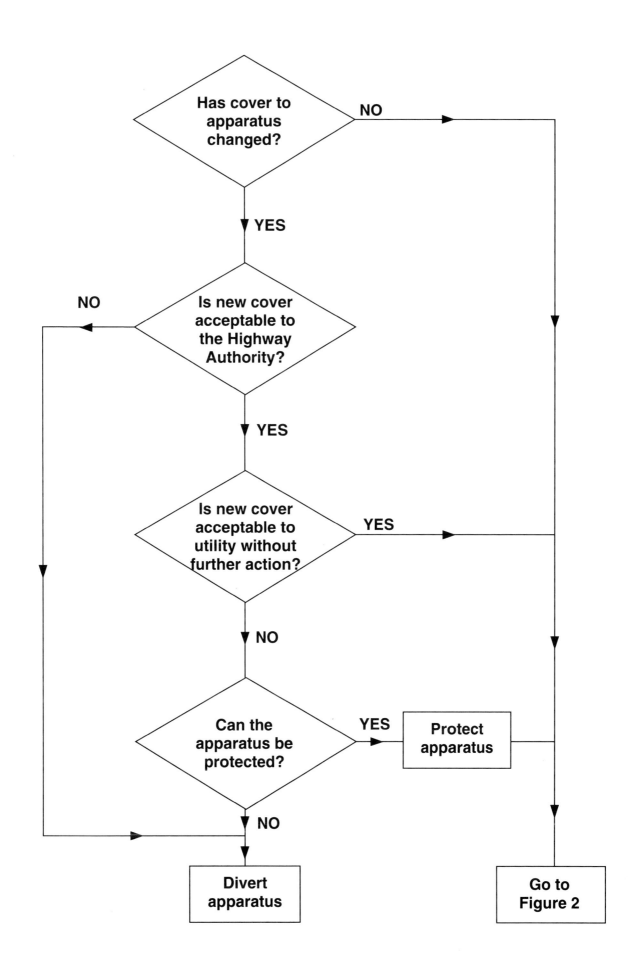

FIGURE 1 CHANGE IN DEPTH

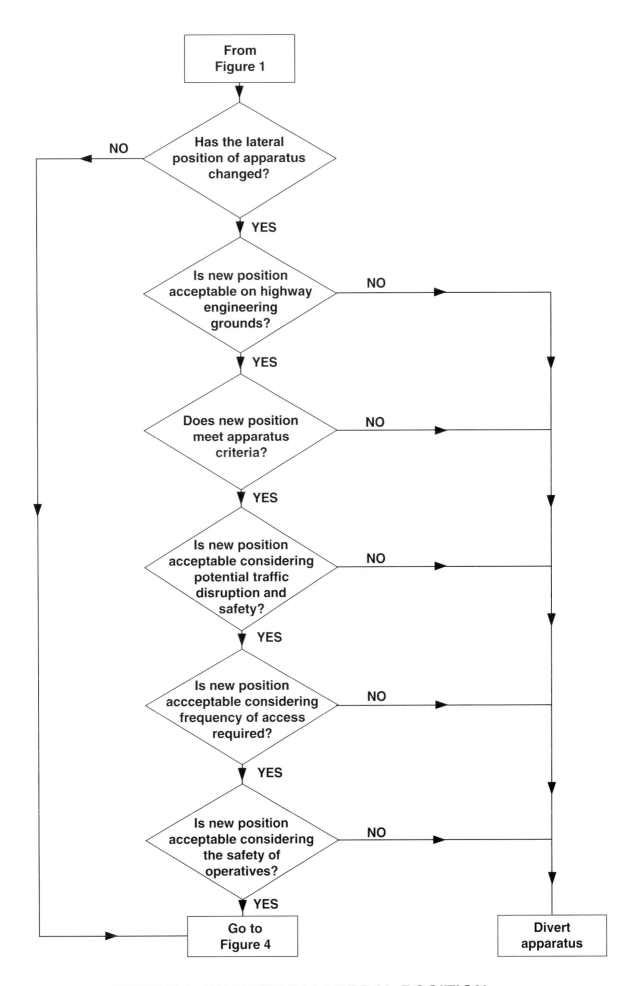

FIGURE 2 CHANGE IN LATERAL POSITION

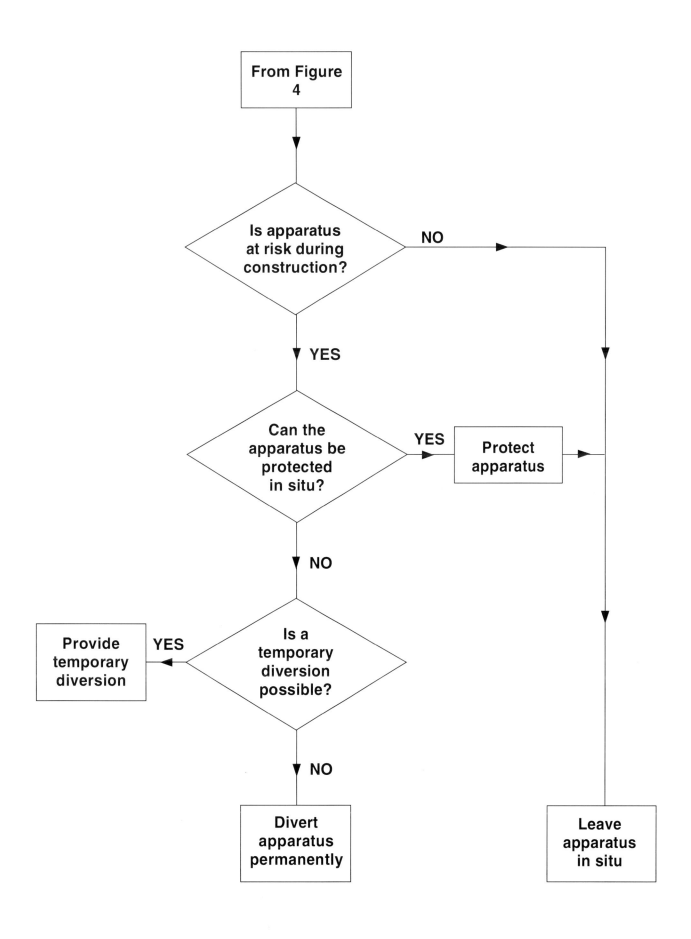

FIGURE 3   APPARATUS AT RISK DURING CONSTRUCTION

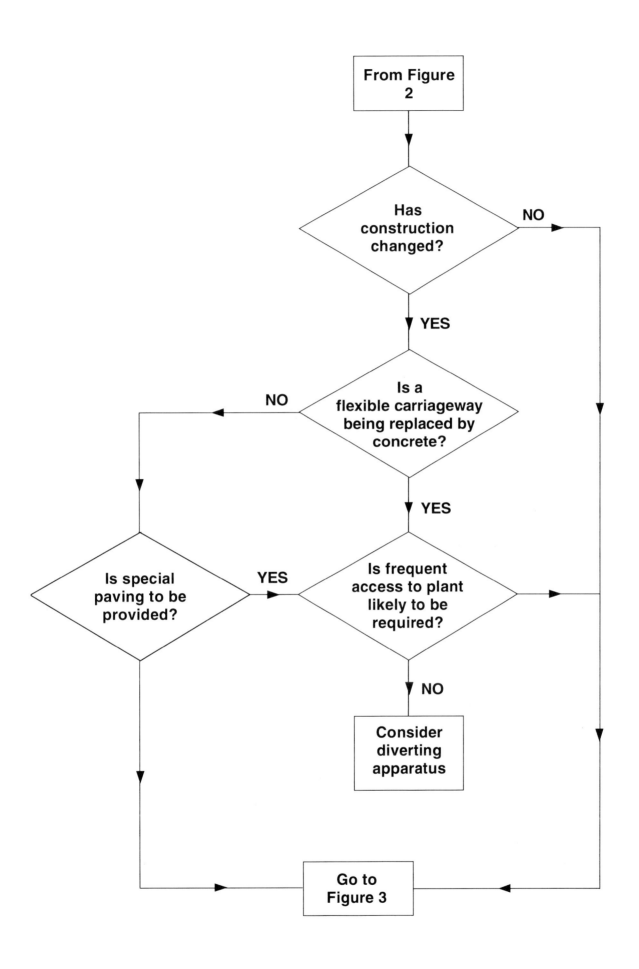

FIGURE 4 CHANGE IN CONSTRUCTION

6 REDUNDANT HIGHWAYS*

6.1 Formal closure (Stopping Up)

The term 'redundant highway*' applies to a section of highway* which the highway authority* or Secretary of State wish to close under their formal legal powers, which are described in Appendix D. In effect this means that the land, including layby, verge or footway, is no longer required for any highway* purpose. Historically most highways* were constructed over private land, the highway authority only having the right to provide and maintain the surface. Hence if a highway* is closed, (and in Scotland it is no longer used as a road) the rights of passage are extinguished and the land ownership normally reverts to the original owners, usually considered to be the frontagers.

The highway authority* would normally seek a stopping up order when they wish to remove the right of public access to the land and hence remove their own duties with regard to maintenance and public liability. The closing of the redundant highway* would enable it to be physically separated from the adjacent public highway*, avoiding unauthorised activities such as fly tipping or encampments, and bringing about a significant environmental improvement.

When a highway* which is to be the subject of a stopping up order contains undertakers' apparatus, the highway authority* should be aware of the undertaker's need for adequate access or protection, and should discuss the intended closure with him at an early stage. In the interest of achieving the least cost solution, the general presumption should be that the apparatus will remain in position where there is no need to disturb it, even though there may be some detriment to the owner.

The rights of undertakers are protected under the various stopping up orders as explained in Appendix D. The highway authority* should inform undertakers of the proposed stopping up and the undertakers should then ensure that the highway authority* are aware of his requirements so that provisions can be made either in the stopping up order or by agreement. If an undertaker's rights are adequately protected, and the route continues to be suitable, he will not be expected to object to the stopping up order. The undertaker should also consider the provision of ducts in the new or improved section of highway* to facilitate a future diversion. If the highway authority* installs ducts as part of its works then ducts should be free-issued by the undertaker.

Highways* may also be closed to facilitate development. In this case it would be normal for the developer to negotiate with the undertaker and the highway authority* for a diversion or a wayleave or easement.

6.2 Change of Use

In addition to redundant highways* there are instances where parts of a highway* become surplus to their original purpose (i.e. as a carriageway or footway or verge) but as the result of improvement could now be used as a layby, for example. If no formal stopping up order is made, then this surplus land still retains the legal status of a highway* and undertakers' rights are protected.

Where a highway authority* wish to use part of the original highway* (e.g. a layby) for the temporary storage of materials, care should be taken to avoid obstruction of access to any undertakers' apparatus which is present. It should also be noted that salt stockpiles can contaminate the ground and accelerate the corrosion of certain apparatus.

7 MAJOR BRIDGE WORKS AND BRIDGE REPLACEMENT SCHEMES

7.1 General

Lengths of highway* carried on bridge structures differ from those carried on a natural formation. The differences are -

   i.   construction depth is restricted and relatively shallow; and

   ii. width is similarly restricted and relatively expensive to increase.

7.2 Major Bridge Works

Minor maintenance and repair of bridges can normally be undertaken without affecting any apparatus laid in the highway* over the bridge. 'Major bridge works' mean, by virtue of section 88 of the Act (for England and Wales) and section 147 of the Act (for Scotland), works for the replacement, reconstruction, or substantial alteration of a bridge.

Examples of such works include -

   i.   Reconstruction, which may include strengthening, of a bridge.

   ii.   Alteration, such as adding a new span or lifting or widening the deck of a bridge.

   iii. Waterproofing the deck of a bridge.

When major bridge works are to be undertaken on a bridge carrying undertakers' apparatus, that apparatus is almost inevitably affected. The guidance as to planning and liaison for major highway works* in section 4 above applies equally to major bridge works.

7.3 Methods of Undertaking Major Bridge Works

The method to be adopted in undertaking major work on a bridge is largely determined by the measures to be taken to deal with the traffic using it. The principal alternatives are -

   i.   divert traffic from the bridge on to a temporary structure alongside or to an alternative route;

   ii.   close the bridge for a short period; or

   iii. divide the street* carried by the bridge longitudinally and undertake the work in two or more sections, restricting traffic to an unaffected section.

Alternatively, the intention may be to rebuild the bridge on a new site.

7.4 Methods of Dealing with Undertakers' Apparatus

Whichever method is adopted to undertake major bridge works, or during bridge replacement, any apparatus carried by the bridge will require attention. Consultation between undertakers and bridge owners should take place as early as possible during the design stage to identify the optimum solution. This consultation should take account of the principles in section 2 above. The options available include the following:

   i.   Support and protection of the apparatus on existing alignment during construction. (See also section 5.4 above). This is more likely to be possible with the apparatus of the cable undertakers.

   ii.   Disconnection of the apparatus, either permanently or temporarily. Permanent disconnection will rarely be feasible. The scope for temporary disconnection will depend on the duration of the bridge work and alternative means of ensuring continuity of supplies.

   iii. A two-stage diversion of the apparatus either on to the unaffected section of the existing bridge or on to a temporary structure. On completion of the bridge works, the network is re-established by connecting to existing or new apparatus in the reconstructed or altered bridge.

    iv.   A single-stage diversion of the apparatus to a permanent structure which may be a new highway* bridge or a special undertakers' bridge, the latter option may require a separate agreement. Alternatively the diversion could be to a new position, e.g. under the river, railway track or canal crossed by the highway* bridge.

Protection or diversion may also be necessary for apparatus laid alongside foundations, piers or abutments when these require repair or alteration.

As far as is practical, undertakers' apparatus should be located away from bridge parapets and arch spandrels. It may be possible to incorporate pipe bays or ducts in bridge decks to accommodate existing apparatus. Allowance should also be made in these for undertakers' increased and/or future requirements, subject to appropriate financial contributions (see also section 9 below).

8 REDUNDANT BRIDGES

There are two types of case to consider:

i. The highway* carried by the bridge becomes redundant.

   This situation usually arises where a replacement bridge has been built nearby, enabling the old highway* to be closed. If the redundant bridge is to be demolished, arrangements should be made to divert undertakers' apparatus into the new bridge or to an alternative route. If the bridge owner decides to retain the redundant bridge, any existing undertakers' apparatus can generally remain. However, new apparatus would not normally be placed in a redundant bridge.

ii. The feature, e.g. railway or canal, which the bridge spans becomes disused but the highway* is maintained.

   In these cases, the bridge may be retained on a permanent basis, with or without infilling the feature, and apparatus carried by the bridge can remain. Any apparatus passing under the bridge is likely to have to be diverted. Undertakers should endeavour, wherever practical to avoid placing new apparatus in a bridge which is to be demolished. Undertakers and the authority should co-operate to find an alternative location for such apparatus.

9 GENERAL STATEMENT OF FINANCIAL AGREEMENT

9.1 General

Section 85 of the Act (for England and Wales) and section 144 of the Act (for Scotland) provides for Regulations prescribing the way in which the allowable costs of diversionary works will be shared between relevant authorities and undertakers. These are contained in the Street Works (Sharing of Costs of Works) Regulations, SI 1992 No. 1690 and the Road Works (Sharing of Costs of Works) (Scotland) Regulations, SI 1992 No 1672 (S.159).

This section sets out the agreed cost sharing principles. Detailed arrangements for their implementation are set out in Appendix C section C9.

9.2 The Standard Cost Sharing Principle

The authority will bear 82%, and the undertaker 18% of the "allowable costs" of diverting or protecting apparatus required as a result of major highway*, major bridge and major transport works - "diversionary works" - **provided** the authority meets the payment schedules as set out in Appendix C, section C9.2 and C9.3.

The "allowable costs" are the costs of the works described in Appendix C section C4 and do not include financing charges, nor the costs of either party in respect of that part of the work described in Appendix C sections C2, C3 and C4 which consists of preliminary planning and liaison.

9.3 Variations on the Standard Principle

9.3.1 Change of Carriageway Construction

The full cost of diversionary works will be met by the undertaker where they arise solely from a change in carriageway construction in the circumstances set out in section 5.5.(ii) and (iii) above and the authority has not required the apparatus concerned to be diverted.

9.3.2 Redundant Highways* and Bridges (see sections 6 and 8 above)

There are three possible cases:

> Case a.
> **Where apparatus is in a highway* or bridge prior to the declaration of redundancy** - the standard cost sharing principle described in section 9.2 above applies to the relocation of that apparatus and any replacement or renewal of it.
>
> Case b.
> **Where apparatus is placed in a highway* or bridge (other than by way of renewal or replacement) after the declaration of redundancy** - provided the highway* becomes redundant within five years, or the bridge within ten years, of the declaration of redundancy, the allowable cost of relocating that apparatus will be met in full by the undertaker.
>
> Case c.
> **Where apparatus is placed in a highway* or bridge after the declaration of redundancy, but the highway* or bridge does not become redundant within five or ten years respectively** - this is the same as Case a. above.

9.3.3 Replacement or Modification of a Bridge

The design and construction of a replacement or modified bridge should take account of the need to accommodate undertakers' apparatus equivalent in size and mass to that which is in the existing bridge, e.g. by providing sufficient space and structural strength. The costs arising from these requirements will be borne by the authority. However, the cost sharing principle described in section 9.2 will apply to the allowable costs of housing or supporting the apparatus, e.g. ducts, pipe bays or hangers, and the diversion of the undertaker's apparatus.

If an undertaker wishes to take the opportunity of the bridge works to increase the provision for its apparatus in the bridge, it must meet the additional cost involved.

9.3.4 Special Case of Change of Depth

The Code gives advice on the diversion of apparatus due to a change of depth (section 5.2 above and Appendix B). The standard cost sharing principle will apply in most cases. However, there may be some cases, as follows, where different financial arrangements may be needed:

    a. **where apparatus is at sub-standard depth and outside the limits set out in this Code (even for protection).**

        In this case, the authority and the undertaker should consider whether there is justification for the undertaker to make a greater contribution to the cost of diversionary works. An increased contribution would not be expected where the undertaker has already accepted reduced depth as a result of previous highway* improvement.

    b. **where the undertaker has already made provision for future highway works* which are subsequently varied.**

        In this case, it is not likely to be appropriate to expect the undertaker to make the 18% reduction in cost.

9.3.5 Works Funded Wholly or Partly by a Private Developer

Where the authority is entitled to receive a contribution from a private developer, in cash or kind, towards the cost of diversionary works (whether or not the contribution is received) or itself is the developer and has decided to make such a contribution from non-highway* funds, the 18% allowance will only apply to that part (if any) of the cost of diversionary works not covered by the contribution. The developer will not get the benefit of the cost sharing arrangements.

It is common practice for developers to be required to pay for undertakers' works in full before such works are put in hand. This is to protect the undertakers should the scheme not proceed after the undertakers' works have been carried out.

Section 85 of the Act (for England and Wales) and section 144 of the Act (for Scotland) enable the authority, on behalf of the undertaker, to recover from a third party (for example, a private developer) costs incurred in carrying out diversionary works and the authority is expected to do so in all appropriate cases.

10 CONCILIATION AND ARBITRATION

10.1 Introduction

It is an intention that this Code will provide sufficiently detailed guidance so that agreement on the nature and extent of any necessary work is reached at local level. Organisations at the local level should always use their best endeavours to achieve a solution to any issue without having to refer the matter to conciliation. This might be achieved by referring the issue to senior management for settlement. If however, agreement cannot be reached, then the following procedure should be invoked.

10.2 Conciliation

If discussions at local level fail then in order to resolve any issue as quickly and informally as possible an attempt should be made at conciliation in the following manner -

i. An independent conciliator acceptable to both parties shall be appointed from a panel of local undertaker and local authority "senior officers" and nationally agreed consultants. The decision as to whether or not to proceed with conciliation should be determined through the appropriate undertaker or local authority procedures, but should be taken at a senior level and must be reported to the Highway Authorities and Utilities Committee (HAUC).

ii. Wherever possible the conciliation should be undertaken by one person. However, a conciliation panel of three persons may be set up where it would be unreasonable to expect one person fully to encompass the technical matters involved. In such a circumstance each side of the dispute shall nominate one member of the panel and the chairman shall be a mutually acceptable independent person.

iii. Each party shall bear its own cost of conciliation, and any fees or expenses of the conciliation panel shall be borne equally by the parties to the dispute.

iv. Each party at its own cost shall make available to the conciliator or panel such technical and costing information as he or they may require to facilitate a fair and reasonable decision to be made.

v. The conciliation should take place within two months from the date on which the issue is referred to HAUC.

vi. The panel of conciliators should be agreed at the first meeting of HAUC in each calendar year or at such other time as HAUC agree.

10.3 Arbitration

By virtue of section 84(3) of the Act (for England and Wales) and section 143(3) of the Act (for Scotland), any dispute between the relevant authority and the undertaker as to the identification of necessary measures, settling of a specification, and coordination of those measures and the authority's works etc, shall, in default of agreement, be settled by arbitration. Where the conciliation procedure has not resolved the issue or the conciliation procedure has not been used, the issue shall be referred to an arbitrator appointed by agreement or, in default of agreement, by the President of the Institution of Civil Engineers (in England and Wales) or by the Sheriff (in Scotland) - see sections 99 or 158 respectively of the Act. Arbitration procedures should be instigated within two months of the conciliation report being issued.

NOTE:

References in this section to HAUC should be taken to refer to Regional HAUCs unless an issue of major principle arises which may be appropriate to be dealt with at national HAUC level.

APPENDIX A

TYPES OF UNDERTAKERS' APPARATUS

This appendix provides an outline description of the network and apparatus of each of the main undertaker industries. The intention is to give sufficient detail to help the reader understand the terminology used in other parts of the Code and to describe undertakers' practices which impact upon major highway*, bridge and transport works. It must be understood that the apparatus actually in the ground may represent a variety of practices adopted over a long period of time. What is described here refers primarily to present-day practice.

A1   GAS INDUSTRY

Gas mains are classified by pressure level as follows -

   i.   high pressure mains operating at pressures above 7 bar (100 psig)

   ii. intermediate pressure mains operating at pressures between 2 and 7 bar (30-100 psig)

   iii. medium pressure mains operating between 75 mbar and 2 bar (1-30 psig)

   iv. low pressure mains operating at pressures up to 75 mbar (1 psig).

High pressure mains generally run across country and are more likely to be affected by new roads or major highway* alterations, rather than by changes of roads in urban areas or by more minor road realignments, although road crossings are more likely to be encountered. It should be recognised that there are restrictions on the locations of these high pressure mains, and any alteration which is necessary is likely to be expensive and considerable notice will be required for the acquisition of materials. These high pressure mains are constructed in steel.

Materials which have been, or are used, for gas mains are steel, spun or pit cast grey iron, ductile iron, polyethylene and, in a limited way, pvc and asbestos cement. The differing properties of these materials can influence decisions on whether or not it is necessary to carry out diversionary works. New polyethylene mains may have been inserted into abandoned metallic mains.

In addition to mains, there are services which supply individual properties; these are generally connected to low pressure mains, but they may also be connected to medium pressure mains, and in a very limited number of instances, to intermediate or high pressure mains.

There is also ancillary equipment to consider, such as valves, syphons, pressure points and pressure regulating equipment. Pressure-regulating equipment may be located in above-ground buildings or kiosks, or located in below-ground chambers or modules.

Small diameter piping is usually associated with this ancillary equipment, and this often runs from the main to surface boxes adjacent to that ancillary equipment.

It is very important that gas supply systems remain charged at pressure for reasons of safety. If consumers' supplies are interrupted, for instance from interference damage to gas apparatus, individual consumers have to be visited, firstly to make the position safe, and secondly, after repair of the apparatus, to purge and restore individual supplies. Apart from the obvious danger from damage to gas apparatus, the cost implications are therefore very likely to extend beyond the repair of the immediate damage.

Depending on the characteristics of the ground, various measures are nowadays adopted to prevent corrosion of metal pipes. Ductile iron pipes may be protected by loose sleeves of sheet polythene. Steel pipes are coated and may have cathodic protection applied from the public power supply or by sacrificial anodes. Anodic ground beds are situated at intervals along the main. Corrosion takes place rapidly at any point of damage to the coating or sleeving, and it is important that any damage is notified to the gas undertaker so that it may be properly repaired.

Cast or ductile mains operating at medium pressure may have thrust blocks, usually made of concrete, located at bends and at other fittings.

The inherent nature of the product being transported is an important factor when considering requirements for alterations and is very important when considering measures for safeguarding apparatus during the construction phase.

A2 WATER INDUSTRY

A2.1 Water mains

Distribution mains are those mains from which service connections are taken to supply customers' premises. They are usually of less than 300 mm diameter. The service connection pipes, known as communication pipes, extend from the main to the street* boundary and a stopcock is usually positioned near the boundary and usually marks the limit of water undertaker's responsibility for the pipes.

Trunk mains are feeders which have few, if any, service connections from them.

Generally, mains carry treated water suitable for domestic consumption - such mains have specific water quality requirements which have to be maintained. There are also mains carrying untreated raw water between points of abstraction and treatment works.

Materials which have been, or are, used for water mains in streets are steel, cast iron, spun iron, ductile iron, asbestos cement, uPVC, polyethylene and, occasionally, prestressed concrete and GRP. The differing properties of these materials can influence decisions on whether it is necessary or not to carry out diversionary works.

Materials for service connections are, or have been, galvanised steel, lead, copper and polyethylene.

Special bedding and surrounding materials may be needed for pipes, especially for plastic pipes, and pipes laid in contaminated ground.

The normal minimum depth of cover for water mains is 900 mm. They are deeper than the other apparatus in the highway* excepting sewers, in order to secure protection from disturbance and traffic loading and for compatibility with the service connection pipes whose minimum depth of 750 mm is necessary to avoid freezing.

Ancillary apparatus on mains includes sluice valves and other forms of shut-off valve, air release valves, pressure and flow regulating valves, fire and washout hydrants, meters and in-line boosters. The bodies of hydrants and manually-operated, ungeared sluice valves are buried and only the upper parts are housed in a brickwork or precast concrete surround to allow them to be operated through a surface box. Other apparatus is installed in underground chambers, usually of brickwork with reinforced concrete roofs. Increasingly such apparatus has power and/or telemetry cable connections to enable it to be operated automatically or by remote control. Above ground, kerbside or boundaryside kiosks are often needed to accommodate power and electronic control gear.

Depending on the characteristics of the ground, various measures are nowadays adopted to prevent corrosion of metal pipes. Spun and ductile iron pipes may be protected by loose sleeves of sheet polyethylene. Steel pipes are usually sheathed in bitumen or plastic and may have cathodic protection applied from the public power supply or by sacrificial anodes. Anodic ground beds are situated at intervals along the main. Corrosion takes place rapidly at any point of damage to the coating or sleeving, and it is important that any damage is notified to the water undertaker so that it may be properly repaired.

Water mains need trench side support for thrust blocks, usually of concrete, which are provided to resist the hydraulic pressures exerted at bends and T-junctions. The larger steel and GRP mains also rely on trench side support to maintain their shape.

A2.2 Sewers

Gravity sewers normally flow part-full. They require manholes at each change of direction and are generally much deeper than other services. Manholes are of brickwork or concrete construction and consist of a working chamber connected to the ground surface with an access shaft. Sewer junctions are made at manholes but most drains take the shortest route and connect to the sewers between manholes.

Pumped sewers flow under pressure, are similar to trunk water mains, but have less ancillary apparatus.

Materials which have been or are used for gravity sewers are brickwork, clayware, concrete, uPVC, GRP and pitch fibre. Drain connections are usually of clayware or uPVC. Pumped sewers may be of steel, cast iron, spun iron, ductile iron, asbestos cement or polyethylene.

As brick sewers rely structurally on full integrity of cross-section, any damage can result in extensive collapse and must be repaired at once.

A3 TELECOMMUNICATION AND CABLE TELEVISION INDUSTRY

This section primarily refers to the long-established BT network, the systems of more recent telecommunications operators will differ in a number of respects but the broad principles are likely to apply.

The telecommunication cable network can be divided into three main sections -

i. Trunk network - long-haul high-capacity systems providing interconnection between main centres of population.

ii. Junction network - interlinks local exchanges or connects a local exchange to a trunk exchange.

iii. Local network - connects the customer's terminal equipment to the local exchange. The local network is sub-divided into -

   (a) Local distribution - from the primary cross-connection points either to distribution points then to the customer by service feeds or else direct to the customer.

   (b) Direct lines - in the case of some major customers, dedicated cables may link the customer direct to the exchange.

In the terminology adopted in this Code, the 'trunk mains' element would encompass the trunk, junction and local main parts of the network, the 'service line' element would refer to the local distribution part.

Telecommunication apparatus liable to be affected by road diversion works may be divided into six categories-

i. **Civil Engineering**

   This category subdivides into duct and jointing chambers -

   (a) **Duct**

   A large percentage of the total underground cable network is contained in duct, directly-buried cable being used in some rural situations and for the feeds to customers' premises (however, such directly-buried cable tends to be replaced by a ducted system for new work). Most of the older duct is earthenware although for new work plastic duct only is used. Various other materials have been used in relatively small quantities. Except at the extremities of the distribution network ducts are likely to be installed in multi-bore tracks, sometimes in city centres containing in excess of fifty bores. Tracks with a small number of bores generally have the ducts laid directly in the ground in soft bedding without the additional concrete used to give stability to larger nests of ducts. The great advantage of a ducted cable system is that additional cables can be installed, old cables recovered or maintenance carried out from jointing chambers without having to excavate. Excavation is then limited to the provision of new ducts or clearing blockages in bores.

   Wherever practical, ducts are laid under the footway or verge, a practice reinforced by licences issued under the Telecommunications Act 1984.

(b) **Jointing Chambers**

Jointing chambers are of two broad types - the manhole, which is an underground chamber accessed by a shaft from the surface, and the joint box, which is a relatively shallow depth box with the whole of its top surface consisting of removable covers. Access covers for jointing chambers are available for different duty applications and used as appropriate in the carriageway, driveway or footway settings. Manholes are usually constructed in reinforced concrete although some exist which have been built in brick. Joint boxes may be constructed in reinforced concrete, brick, or, in the case of the smallest sizes, prefabricated glass-reinforced plastic or high density polyethylene.

ii. **Cables**

(a) **Cable Types**

A wide range of cable types with copper or aluminium conductors or optical fibres is in use. The most common types are the twisted-pair or quad-type cables consisting of fine-gauge copper or aluminium wires. Such cables may contain from two pairs up to 4800 pairs of wires. Coaxial cables are in use where high-capacity, high quality circuits are required although the use of such cables is decreasing as their functions can be better served by optical fibre cables. Optical fibre cables are used extensively for trunk and junction systems, and will increasingly replace the use of metallic conductor cables in the local networks.

Nearly all modern cables are sheathed in polyethylene but older cables have lead sheaths. To exclude moisture from many major cables they are pressurised with dry air. Other cables are usually jelly-filled for the same purpose.

(b) **Jointing**

The wires in cables of the pair or quad type can be jointed by a simple operation, often partly mechanised, although the operation will have to be repeated many hundreds or thousands of times for the larger cables. In the case of coaxial cables, the jointing of the individual coaxial tubes is an operation requiring precision and skill. In both wire and coaxial type cables, the joint itself does not significantly degrade the overall system performance and additional joints introduced as a result of highway works* are normally acceptable. This is not the case with optical fibre cables as splices in the fibre introduce significant loss of signal. It is therefore important to minimise the total number of splices in a system.

(c) **Sheath Closures**

Telecommunication cables are very susceptible to damage should there be ingress of moisture and special precautions are necessary to maintain cables in a dry condition. Ingress of moisture is most likely at sheath closures. Older types of closure in particular are sensitive to disturbance and any movement resulting from cable alterations necessitated by highway works* is likely to require the sheath closures to be remade.

iii. **Electronic Equipment**

On high-capacity cable systems it is often necessary to provide electronic equipment at intervals along the cable route. Such equipment may be housed in underground jointing chambers or in roadside above- ground cabinets. If underground, such equipment is normally not under the carriageway to minimise problems resulting from traffic-induced vibration and to facilitate access for maintenance.

iv. **Cabinets**

Cross-connection points are housed in above-ground cabinets and frequent access to these cabinets is necessary for reconfiguring the cable system and for maintenance. It must be possible to open cabinet doors and undertake work inside the cabinets without endangering staff or road users.

v. **Overhead Apparatus**

Poles perform one of two functions. Firstly, they are used as distribution points where an underground or overhead cable feed to the pole is terminated and small individual cables radiate from the polehead to serve customers' premises. Such poles are frequently climbed. The other main application for poles is for overhead cable routes. Poles subjected to unbalanced cable loadings will require staying or strutting.

Many "joint user" poles exist. These are usually where telecommunication cables are attached to electricity supply poles, but sometimes power cables are attached to telecommunication poles.

Public Call Office

Call offices are generally sited in the footway and telecommunications and electricity cables are led underground into the call office.

A4 ELECTRICITY SUPPLY INDUSTRY

The electricity network can be divided into two sections -

A4.1 The Transmission Network

The transmission network (often referred to as the "grid") forms the major interconnection of power stations and operates at voltages of 275 kV and 400 kV. Distribution system supplies, described below, are taken from major substations operating at 275 kV/132 kV or 400 kV/132 kV. This network consists mainly of overhead lines supported by towers but at certain locations underground cable has to be used and, where this is situated within the highway*, it would normally be sited beneath the carriageway. These cables are more likely to be affected by new roads or major highway* alterations rather than by changes in road construction in urban areas or by minor road realignments.

Alterations to these circuits are invariably expensive and require considerable notice for the acquisition of materials and operational planning.

A4.2 The Distribution Network

This can be classified by voltage pressure as follows -

i. low voltage services supplying individual domestic or small commercial/industrial consumers normally at 240/415 volts

ii. low and medium voltage mains operating at less than 1000 volts

iii. high voltage mains operating from 1 kV to 22 kV

iv. extra high voltage mains operating from 22 kV to 132 kV

In rural areas, the distribution network is generally constructed of overhead lines whilst in urban areas underground cables are usually employed.

Apparatus associated with overhead lines includes supports, stays, conductors and earth wires. These can be affected by highway* modifications as follows -

(a) Supports and stays may be undermined causing instability

(b) The ground clearances of overhead conductors may be affected where road levels are changed

(c) Earth wires are placed on supports and also within the ground for the protection of the apparatus, customers and the general public. Care must be exercised when operating in proximity of any overhead apparatus.

Certain customers requiring large supplies draw their power direct from high voltage systems. The costs of diverting plant operating at the higher voltage levels are greater than an equivalent diversion of lower voltage plant. Extra notice time, for material acquisition and operational planning, is normally required when higher voltage plant is diverted.

NB The definitions of voltages given above are the most likely reference to be found on electricity companies' plans of cable routes. The current international definitions of voltage levels are not the same.

A4.3 General

The underground mains of some undertakers are often placed inside pipes or ducts. However this is not generally acceptable for electricity power cables since they need to be in direct contact with the soil in order to dissipate the heat generated by electrical losses. Earthenware or plastic ducts are used however at road crossings, bridges and other similar short distances to avoid traffic disturbance and to ease the future replacement of faulty cables.

For many years, underground cables of all voltages were mainly constructed of paper insulation, with copper or aluminium conductors and a protective lead sheath. Steel wire or steel tape armour was commonly used with bituminised hessian overall covering. For at least 10 years, LV and 11 kV cables with aluminium sheaths, and PVC over-sheaths have been used as alternative protection. Cables with polymeric insulation, aluminium conductors and an extruded polymeric sheath have also been used for certain voltages. Various combinations of materials and designs may be employed depending upon circumstances.

At the higher voltages (ie. 25 kV and above) gas-pressurised and oil-filled cables are used requiring pressure tanks, link boxes and other auxiliary equipment. Polymeric cable over-sheath is sometimes used acting as an anti-corrosion barrier to protect the internal pressure-retaining sheath. Gas compression cables are also used and this type of cable may be difficult to distinguish from a steel water or gas pipe. The increased probability of the pressure-retaining sheaths being damaged as the result of highway works* can influence decisions on whether or not the cable needs to be diverted.

Additional protection of cables from mechanical damage is normally provided where necessary by the positioning of inscribed concrete slabs, earthenware tiles, plastic tiles and tough plastic tapes. Marker tapes may be placed over cables to warn excavators.

Cable joints are required at all voltages to either connect the cable sections together or to service adjoining consumers. Joints are vulnerable items in the cable network and need additional protection. This results in the higher voltage cables having physically larger joints with consequently larger excavations and longer times for undertaking the work. Low voltage cable networks may be provided with disconnection facilities mainly accommodated in accessible cast iron joints placed in brick pits beneath manholes.

When diversionary work is to take place it is usually essential to maintain the existing network whilst installing new plant since existing customers cannot be disconnected for long periods. Plant operating at the higher voltages is site-tested before it is connected to the existing network.

The unique nature of the product being distributed is an important factor when considering requests for alterations or measures to safeguard the plant during the construction phase.

A4.4 Categories of Cables and Apparatus

These are as follows -

(1) Low voltage service cables supplying domestic or small commercial consumers

(2) Low voltage mains cables (LV) - normally 415/240 V - to which service cables are jointed at irregular intervals

(3) High voltage cables (HV) - 6.6 kV or 11 kV

(4) Extra high voltage cables (EHV) - usually 33 kV or 66 kV (some cables are pressure-assisted with nitrogen or oil)

(5) Supertension underground cables 132 kV, 275 kV and 400 kV of various types. (some are pressure-assisted and some are water-cooled with associated cooling equipment).

(6) Auxiliary equipment such as link boxes, feeder pillars, control kiosks (for pressure-assisted cables), pressure tanks, earth wires, oil pipelines and cooling pipes

(7) Duct or pipe systems, cable draw pits or chambers, expansion chambers which may contain any of items (1) to (5)

(8) Buried transformers and underground or above ground substations with ventilation cabinets and LV distribution pillars.

(9) Apparatus associated with overhead lines, such as supports, stays, conductors and earth wires. Any of items (1) to (6) above which may be laid underground from certain overhead line supports.

(10) Telecommunications apparatus of the electricity companies is generally laid independently of electricity cables and the considerations appropriate to telecommunications apparatus in the code apply.

(11) Buried pilot wires for control purposes may be associated with underground or overhead cables.

NB Items (1), (2), (3) and (4) are usually (where possible) laid in footpaths. Item (5) is normally laid in carriageways but occasionally in footways, depending on the circumstances.

APPENDIX B

UNDERTAKERS' APPARATUS - FACTORS MATERIAL TO HIGHWAY WORKS*

This appendix outlines the constraints resulting from the nature of undertakers' apparatus and the use to which it is put which have a direct impact on the decisions whether or not to divert or alternatively, to protect or strengthen the apparatus.

B1 GAS APPARATUS

B1.1 General

There are a number of general points in relation to gas apparatus which will affect decisions on diversions or protection. These include the -

    i.   need to maintain continuity of supply

    ii.   inherent nature of the product being distributed

    iii.   operating pressure of the apparatus

    iv.   duty of the apparatus

    v.   material of the apparatus

    vi.   type of construction of the apparatus.

When diversionary work is to take place, it is normally essential to maintain the existing system and to lay new apparatus in the diverted position. The new apparatus will then be connected into the existing system. These connections have to be done under gas pressure, and sometimes rider arrangements are also necessary. The costs of these connections can therefore be considerable and on short diversions they often make up a substantial proportion of the total cost of the works. Working under gas pressure on high and intermediate pressure mains is extremely expensive.

After the new apparatus is charged with gas, it may be necessary to maintain gas within the old apparatus until service connections are transferred to the new main, thus allowing the old apparatus to be taken out of commission. Given the cost of connection work, it may be cheaper to lay additional pipe in the diversion if this can reduce the number of connections.

There may be restrictions as to the time of year when connections can be made. These restrictions are more likely to apply to high, intermediate and medium pressure mains.

B1.2 Depth of Cover

The normal minimum cover for mains operating within the low and medium pressure ranges in footways or verges is 0.6 m, and in carriageways is 0.75 m, although some older mains may have been laid at different depths. High pressure mains are generally laid at a depth of cover of 1.1 m.

It may be acceptable to reduce the cover of medium or low pressure steel mains for short sections, although additional protective measures might be necessary.

Maximum cover which would affect the pipe or joint will vary depending on the material, but for simple guidance it would be unusual to accept covers greater than 1.2 m for other than very short sections. It should also be recognised that increased cover is less acceptable where access is likely to be more frequent, for example on low pressure mains.

In all cases where it may be possible to leave the main in situ (increase or decrease in depth) trial holes are likely to be made to ascertain the structural condition of the pipe and assess its strength in relation to the changed loading.

Services are laid shallower than mains; the normal minimum depth of cover in carriageways is 0.45 m.

B1.3 Lateral Position

B1.3.1 Change of Location Within Carriageway

It is necessary to consider the pressure level of the main, the material and type of construction and the extent of the highway works*. The effect on ancillary apparatus such as pressure regulating equipment and associated housing and valves, where quite frequent access is necessary, will also have to be assessed:

i. **High Pressure Mains**

There are restrictions relating to property proximity and density on the location of mains operating within this pressure range and on pipelines running within highway* boundaries. The effect on these mains of any road proposals would need very careful individual consideration, although the number of instances that these mains are situated in existing roads (with the exception of road crossings) are limited. It is vital to stress the importance of safeguarding the main during the construction phase and to note that any diversions which are necessary are likely to be expensive and that material acquisition is likely to be lengthy.

ii. **Intermediate Pressure Mains**

The need for frequent access is unlikely and it may be acceptable for the existing position to be maintained. There are however certain restrictions on the locations of mains operating at this pressure which would need to be considered, eg. there are restrictions relating to property proximity and density.

iii. **Medium Pressure Mains**

The extent of plant alteration is likely to be influenced by the type of highway works*, and the material and type of construction of the main.

(a) The need to alter steel or polyethylene medium pressure mains is likely to be limited as long as the key factors of depth etc., can be satisfied.

(b) It will be necessary more frequently to alter cast iron and ductile iron mains where the effect of traffic could increase either during or after the construction phase, due to the material and type of construction.

iv. **Low Pressure Mains**

Similar considerations to those for medium pressure mains apply, although more frequent access will be necessary.

B1.3.2 Change of Location from Footway to Carriageway

It will generally be necessary to move the main after consideration of a number of the key factors.

In addition, it will not generally be possible to allow cast iron mains previously in the footway or verge to be subjected to vehicular traffic loading even if the depth consideration can be overcome.

In certain circumstances it may be acceptable to allow the main to remain in the edge of the road, particularly for short sections such as laybys, road visibility splays and car parking strips. However, such decisions will depend upon the operation pressure, material, type of construction and depth of the pipe. Where decisions are made to leave the pipe these may depend on the provision of additional protection.

B1.4 Risk During Construction

All undertakers have obligations to their customers and need to take reasonable steps to ensure the continuity of supply. In addition, the inherent nature of the product being transported is an important factor when considering requirements for safeguarding apparatus during the construction phase.

The material and type of construction of the pipe system are also extremely important and the extent of the consequences of damage to the higher pressure apparatus, and the vulnerability of some of the older materials used in low and medium pressure pipe construction, particularly cast iron, are key factors.

It should be recognised that existing gas mains cannot be raised, lowered or moved laterally even for a few millimetres. New pipes are necessary for locating apparatus in a new position.

Gas apparatus should not be undermined and certain apparatus is particularly vulnerable to deep excavations adjacent to the apparatus, and proposals to dig deep trenches may mean that gas apparatus will have to be diverted.

B1.5 Change in Type of Road Construction

It would only generally be necessary to consider diversion of apparatus if the type of road construction was being enhanced considerably, which would add significantly to the time involved in gaining access and the cost of doing so, for example the use of reinforced concrete would be a case where diversion may be necessary.

B2 WATER APPARATUS

B2.1 General

Decisions on diversions or protection are likely to be heavily influenced by considerations of access to the mains themselves for repair purposes and to the ancillary apparatus for maintenance and operation. The need to ensure continuity and quality of supply is also of great importance.

B2.2 Depth of Cover

i. **Trunk Mains**

The minimum cover is normally 900 mm but up to 1.5 m may be acceptable. Covers down to 750 mm and up to 2.0 m may be acceptable for short lengths. Trial holes and other investigations are likely to be needed for an assessment of the structural condition of the pipe where the cover is to be altered.

ii. **Distribution Mains**

The normal cover is 900 mm. The minimum acceptable is 750 mm and generally cover should not exceed 1.2 m. For short lengths, cover up to 1.5 m may be acceptable.

iii. **Service Pipes**

The minimum cover for short lengths is 750 mm. They should not have their accessibility prejudiced by being buried at significantly greater depth, unless it is necessary for frost protection in exceptional locations. The maximum allowed by water bye-laws is 1.35 m.

iv. **Gravity Sewers**

Radical changes in depth of cover may require strengthening of the sewer and reconstruction of manholes. Trial holes and/or a CCTV survey should be carried out to assess the structural condition of the sewer if significant changes in depth of cover are contemplated.

v. **Pumped Sewers**

As for trunk mains in (i) above.

B2.3 Lateral Position

Change of Location Within the Carriageway or from Verge or Footpath to Carriageway

   i.   **Trunk Mains**

   If the depth of cover will be acceptable, consideration has to be given to the main's relative importance in the system, to its structural characteristics and condition, and to the potential disruptive effects on traffic of any leakage and subsequent repair. The need for diversion may be indicated if the new road construction will cause increased difficulty in breaking open and restoring the surface, or if access to ancillary apparatus for maintenance and operation will be adversely affected to an unacceptable extent by increased exposure to traffic. But if only infrequent access is expected to be needed to the main and the ancillary apparatus, this should also be taken into account.

   ii. **Distribution Mains and Service Pipes**

   The frequent access needed to distribution mains and service pipes for connections and repairs, as well as the frequent and safe access needed to valves, hydrants, stopcocks and meter boxes, are often pressing reasons for keeping the mains and the associated apparatus in footways and verges where it is practicable to do so.

   iii. **Gravity Sewers**

   Although it must remain possible to gain access to manholes safely, most have to be entered infrequently and access considerations alone will not usually determine a need for diversionary works.

   iv. **Pumped Sewers**

   The considerations for trunk mains apply also to pumped sewers but less ancillary apparatus is likely to be involved.

B2.4 Risk During Construction

The operation of construction plant and lorries over pipes with temporarily reduced cover can be an unacceptable risk. Diversion may be necessary unless protective measures are practicable. Factors influencing the decision include:

   i.   maintenance of continuity of supply and water quality,

   ii. material types and condition,

   iii. inability to raise, lower or slew water pipes, and

   iv. possible loss of ground support to pipes.

B2.5 Change in Type of Construction

Concrete road construction is liable to make the tracing of leakages unduly troublesome and reinforced concrete is particularly difficult to excavate speedily and reinstate satisfactorily. Diversion may be necessary if such construction is intended.

B3 TELECOMMUNICATION APPARATUS

B3.1 General

This section is based largely on the BT network, being the oldest and with the widest range of apparatus. More recently licensed operators are likely to have systems broadly similar to modern BT apparatus.

B3.1.1 Characteristics of Apparatus

Four characteristics of telecommunication apparatus have a major influence on the treatment of such apparatus when affected by highway works* -

i. As the needs for fluid-tightness and safeguarding of a potentially hazardous product do not apply to telecommunication apparatus, a wider range of solutions are open when alterations are necessary.

ii. The need for the rigorous exclusion of moisture from cables and joints places constraints on what can be done to older cables particularly in respect of movement during works.

iii. As virtually all cable provision, recovery and maintenance is undertaken from jointing chambers, the siting of their entrances is especially important. Vehicles used in conjunction with cabling and testing must be able to be located near a jointing chamber.

iv. The ducted network uses identical materials and forms of construction regardless of the value or importance of the cables contained within the ducts.

B3.1.2 Relative Costs of Civil and Apparatus Works

In most circumstances, civil engineering work is less costly than cabling and jointing operations. This is especially so when many cables are contained in one duct track and/or when high value cables are installed.

The cost differential is also apparent when temporary diversions are intended - such diversions are likely to be expensive because jointing work has to be done more than once.

Because of the relatively high cost of cabling work, in a number of circumstances it may be economic to provide a new duct track to be used for all future cable installation work leaving the existing working cables in what has become, as a result of a highway* scheme, a poorly-located position but where ready access can still be assured. In the course of time when all the old cables have been superseded the old duct track will be abandoned. Circumstances where such a solution may be suitable include-

i. An existing duct track becomes located at excessive depth (see section B3.2.3 (i))

ii. The change of lateral position of the duct track results in access to existing jointing chambers becoming very difficult (see section B3.3.1 (iv))

iii. Where the existing ducts are to be moved bodily and there is the risk of ducts becoming damaged in the process and presenting a hazard for future cabling (see section B3.4 (ii)).

B3.2 Depth of Cover

B3.2.1 Preferred Depth of Cover

When installing new ducts in existing roads the depths of cover given below are adopted where possible. The depth quoted is measured from the surface to the top of the uppermost duct or, if encased in concrete, to the top of the concrete block.

Cable Television ducting

- in footway or verge   250 mm
- in carriageway   450 mm

PVC and polyethylene duct - singleway and multi-way up to 20 bores

- in footway or verge   350 mm
- in carriageway   600 mm

PVC duct - over 20 bores

- in footway or verge   600 mm
- in carriageway   900 mm

These depths are based on the need to provide adequate stability to the highway* structure and to accord with the preferred plant layouts recommended by the National Joint Utilities Group. Recommended depths have varied with time and much apparatus will be at other depths, generally shallower rather than deeper. The depth will also vary because of -

i. the need to avoid other buried plant or obstructions

ii. the need to ensure the ducts are at least 75 mm below the underside of the road formation

iii. variations agreed with the highway authority*.

Variations from the preferred values are acceptable within the limits indicated in section B3.2.2 and B3.2.3.

B3.2.2 Reduced Cover

Ducts

For the purpose of long-term structural stability, if a distance of 75 mm or more exists between the top of the uppermost ducts and the underside of the road formation then no alterations will be necessary. However, if during highway works*, heavy plant will be passing over the duct track, the minimum acceptable depth would then be 300 mm unless special precautions are taken as specified in section B3.4 (i). If ducts are below the road formation but the distance is less than 75 mm then a reinforced concrete slab over the existing duct track may be provided. This slab will intrude into the road construction and the use of such a slab would need to be agreed with the highway authority*.

Where the amount of space is very restricted and an adequate depth of cover cannot be provided then steel ducts can be used. Problems in their use restricts their application normally to bridge decks.

B3.2.3 Increased Cover

i. **Ducts**

Where the change in road level will lead to an increase in depth, this will generally be acceptable provided the cover does not exceed 3 m and the duct bores are known to be in good condition. If these criteria are not satisfied then a new duct track at the normal depth for new cables will have to be provided although it may be possible to leave the existing cables in position (see section B3.1.2).

ii. **Jointing Chambers**

The effect of increasing depth of cover is that joint boxes may have to be replaced by manholes if the floor of the existing joint boxes becomes too deep for the plan size of the box. It is not practical to lay down detailed criteria here because of the wide range of joint boxes in use. However, regardless of other dimensions the floor of a large joint box must not be more than 1.65 m below the road surface. The replacement of a joint box by a manhole may require 10 m or more of duct track either side of the chamber to be lowered so that the ducts enter the chamber in a suitable position for cabling.

B3.3 Lateral Position

B3.3.1 Change of Location Within Carriageway

Considerations of traffic disruption, road safety, frequency of access, and safety of the undertaker's staff predominate. The following specific points are relevant -

i. The local distribution part of the network requires the most frequent access for maintenance and cable provision and wherever possible is located in the verge or footway. Access to other parts of the network is likely to be less frequent and mainly directed to maintenance access to cable pressurisation monitoring points and electronic equipment. It is nevertheless essential that safe access is possible to any jointing chamber. Duct tracks will need much less maintenance attention, and only if duct bores are blocked by damage, ingress of silt or tree roots.

ii. If the jointing chambers remain in the original position with respect to the highway* but the duct track between chambers is in a different position then no action would normally be needed.

iii. If the jointing chamber access is located in a different position within the carriageway then no change would normally be required provided there is adequate space and hardstanding for vehicles and trailers to be safely located at the jointing chamber site for cabling, jointing and testing purposes.

iv. If a jointing chamber would become located where future access will be very difficult then it may be possible for a new duct track to be provided in an acceptable position for future cables. It may be possible to leave existing cables in the old route (see section B3.1.2).

v. Side-shaft entries have been constructed on existing manholes in circumstances where normal access becomes impractical. Such shafts have limitations in their use. The walls of the manhole may already be supporting cables, space may not be available on the verge or footway to enable an entry to be constructed, other buried plant may be in the way of a shaft. Long lengths of side shafts pose safety problems, and cabling operations may also be impractical.

vi. Siting of jointing chamber openings in the central reservation of dual carriageways can only be considered if there is adequate hardstanding for operational vehicles off the carriageway, any safety fencing or barriers can be removed to facilitate access, and normal arrangements for the safety of operatives and traffic can be achieved.

vii. In some circumstances where frequent access is required to service parts of the network and occasional access required to other parts, it may be possible to extend some cables or pipes via ducts across the road to a new jointing chamber situated off the carriageway so that the equipment needing frequent access is extended to the new chamber.

viii. Jointing chamber entries must not be located in a drainage channel and must be avoided if practicable at road junctions. They cannot be permitted at hazardous bends, or where work at the chamber entry would prevent entry to service roads or access ways. It is necessary to ensure that sufficient clear space is available in the immediate vicinity of the chamber entry to permit safe removal of heavy covers.

B3.3.2 Change of Location from Footway to Carriageway

As it is standard practice to bury ducts at a shallower depth in the footway or verge, the factor which will most influence the decision to divert in this circumstance is depth of cover. An additional consideration is that a high proportion of the "service" elements of the network will be located in the footway and the undertaker's staffing, practices and equipment will be arranged accordingly. As a result, questions of frequency of access and staff safety are likely to be especially important factors.

The following apparatus considerations also apply -

i. Most joint boxes originally in the footway/verge will require total or partial reconstruction to withstand the increased imposed loading and cater for the likely changes in surface levels.

ii. Many manholes have been designed to withstand highway* traffic loadings regardless of location. However some of the older manholes in the footway will have been designed only for that location and so would have to be rebuilt if they were to be subjected to vehicular loading.

iii. It is very likely that frames and covers will have to be replaced on any jointing chamber which becomes relocated into the carriageway.

B3.4 Risk During Construction

According to circumstances, the apparatus may be protected in situ, temporarily moved to a new location for the duration of the construction work, or else permanently moved to a new position.

i. **Protection**

Where ducts are not exposed during works they will have to be protected by metal plates or tracks if heavy construction plant will be crossing the line of the ducts. If the authority accepts the option of protection where the undertaker recommends but does not insist on diversion, the authority must provide written evidence of adequate indemnity insurance in case damage should subsequently occur during the works.

ii. **Slewing of Ducts**

In some cases it will be possible to accommodate small temporary or permanent alterations in the line of a duct track by bodily slewing, raising or lowering a nest of ducts with the cables in situ. Such movement of a duct track is likely to be less satisfactory with earthenware than plastic ducts. In general, the ageing of lead sheaths on older cables leads to embrittlement of the sheath and such cables cannot be satisfactorily slewed. There is also some hazard to cable joints in these operations and they may have to be remade. If there is a risk of ducts being damaged during the moving operation, it may be necessary to provide new ducts for future cabling work, leaving the existing cables in the old ducts to be abandoned in due course (see section B3.1.2). Should the cables suffer damage then immediate repairs or replacement will be necessary. Where the extent of duct damage is restricted then standard duct repair techniques can be employed.

B3.5 Change in Type of Construction

If special pavings are proposed for pedestrianisation and similar schemes, it is sometimes possible for the telecommunications operator to provide special covers for jointing chambers which allow paving blocks to be laid in the cover to blend into the general surface paving. If the authority provides covers then these must satisfy the undertaker's operational and safety requirements.

B3.6 Overhead Plant

The location of poles must be chosen to enable the polehead to be accessed by ladder or elevating platform. Poles must also be positioned to minimise the risk of damage from road vehicles, give the minimum inconvenience to pedestrians, and avoid obstructing access to properties. Hollow poles, which do not have to be climbed, can sometimes be provided in situations where ladder access is impractical or unsafe.

Road alterations may involve replacement of poles if the clearance under the cables becomes inadequate. Minimum heights above ground for overhead lines may be specified in the licences of telecommunications operators and are, typically:

at any point over a street*   5.5 m

on bus routes	6.1 m
on designated high load routes	6.5 m

Aerial cables are not normally allowed to cross dual carriageways as adequate clearance cannot be provided without a pole support in the central reservation.

Other constraints on alterations affecting poles are the minimum depth of a pole in the ground and adequate clearance from overhead power lines.

B4 ELECTRICITY CABLES AND ASSOCIATED APPARATUS

B4.1 General

B4.1.1 Underground Cables

The parameters to be taken into account when considering (a) diversions of underground cable circuits, or (b) the alternative provision of additional protection for such circuits when affected by highway works*, can be listed as follows -

i. the need to protect and support potentially hazardous equipment from mechanical impact, damage, strain and vibration during and after the alterations

ii. a requirement to maintain security of supply if alternative circuits are not available

iii. the operating voltage of the apparatus

iv. the adverse effect on current rating if the laying depth is revised or the medium surrounding the cable is altered

v. the siting, exposure to traffic, space requirements and suitability of cable jointing pits which may be required.

B4.1.2 Overhead Lines

When considering highway works* the following points have to be considered -

(a) Supports (towers and poles) and associated stays may have to be relocated

(b) Ground clearance may be affected if road surface levels are changed

(c) Earth wires from the supports may have to be relocated.

(d) Buried pilot wires may be associated with the route of overhead lines.

B4.2 Depth of Cover

The depth at which electricity cables or cable ducts are usually laid in the ground is decided by the need to avoid undue interference or damage. The heat dissipation of electricity cables is affected by the depth and backfill in which they are laid and this in turn can alter their load-carrying capacity.

B4.2.1 Supertension and EHV Cables

Normal depth of cover is 900 mm.

(a) **Depth Increase**

This will adversely affect current ratings, i.e. the power transmission level (see section B4.5.2). Whether diversion is necessary or not is likely to have to be determined on a case-by-case basis.

(b) **Depth Decrease**

If cover was decreased to less than 600 mm from the new surface level, or the distance below the formation level reduced to less than 300 mm in a carriageway, any cables would have to be lowered or, if this was not possible, diverted. It may be possible to avoid disturbance by protecting with thick steel plates.

B4.2.2 HV Mains Cables

Normal depth of cover not less than 600 mm in footway or in carriageways.

(a) **Depth Increase**

Diversion of cables is not necessary unless the depth of cover of a cable is increased in excess of 1.2 m to the new surface level.

(b) **Depth Decrease**

If the cover was decreased to less than 500 mm from the new surface level, or the cover to the underside of the bearing surface reduced below 300 mm in a carriageway, any cables would have to be lowered or, if this was not possible, diverted.

B4.2.3 LV Mains and Service Cables

Normal depth of cover not less than 450 mm in footways or 600 mm in carriageways.

(a) **Depth Increase**

Provided no further service connections or link boxes are anticipated cover may be increased. Cable diversion is not normally necessary unless the depth of cover is increased to a figure in excess of 1.0 m.

(b) **Depth Decrease**

As the standard depth is relatively shallow, no depth decrease can normally be tolerated because of the possibility of the new (thicker) bearing surfaces impinging on cable joints which project above the cable. Diversion is normally necessary in such cases. If not diverted, a minimum cover of 300 mm of suitable compacted backfill below the underside of the bearing surface is required.

B4.3 Lateral Position

B4.3.1 Change of Location Within Carriageway

i. **Supertension and EHV Cables**

(a) **Urban and Rural Roads with Heavy Traffic**

With certain EHV and supertension cables consideration would have to be given to the effects of increased vibration and the need for complete diversion. Such cables cannot normally be accommodated in the footway.

(b) **Urban and Rural Roads with Light Traffic**

Such cables cannot normally be accommodated in the footway. Diversion would not be necessary.

ii. **HV Mains Cables**

(a) **Urban and Rural Roads with Heavy Traffic**

No diversion would normally be necessary.

(b) **Urban and Rural Roads with Light Traffic**

   No diversion would normally be necessary.

iii. **LV Mains and Service Cables**

  (a) **Urban Roads with Heavy Traffic**

   For single carriageway roads the cable would be diverted to the footway if there were service connections (present or future) or existing link boxes. If a road was converted to a dual carriageway and there were service connections (present or future) on both sides of the road, mains would preferably be laid in both footways with the provision of cross-road ducts to enable future mains and services to be easily installed.

  (b) **Rural Roads with Heavy Traffic**

   The cable would not normally be diverted to the footway unless there was the possibility of extensive additional service connections.

   If the road was being converted to dual carriageway, the arrangements as for urban roads in (a) above would apply.

  (c) **Urban and Rural Roads and Light Traffic**

   Probably no diversion required unless there is the possibility of extensive additional service connections.

Note 1

In all cases, consideration has to be given to the protection of the general public and the undertaker's staff during repair work on the apparatus due to its revised position in the carriageway.

Note 2

Where cables are laid in cross-road ducts and the carriageway is widened it is normal practice to extend the duct length with split pipes to align with the new carriageway width.

B4.3.2 Change of Location from Footway to Carriageway

Provided that there is adequate cover -

i. **Supertension and EHV Cables**

   Consideration would have to be given to the effects of increased traffic vibration on pipe connections, cable joints, etc. Such cables cannot normally be moved to the footway. (See section B4.5.2).

ii. **HV Mains Cables**

   Certain unarmoured lead-covered cables would normally be diverted, or carefully protected, to ensure there is no damage from traffic vibration.

   For armoured or aluminium sheathed cables no diversion would normally be necessary unless additional connections have been planned.

iii. **LV Mains and Service Cables**

   The cable would normally be diverted to the footway unless there are exceptional reasons for not doing so.

If a road is converted to dual carriageway and there are service connections (present and future) in both sides of the road, mains should preferably be laid in both footways.

NOTE

Consideration has to be given to the protection of the general public and undertaker's staff during repair work on the apparatus due to its change of position.

B4.4 Risk During Construction

Section 5.4 explains that additional protection or temporary diversion may be necessary to prevent damage to apparatus during the construction stage. The amount of protection or scope of the temporary diversion considered necessary will depend upon the circumstances. The hazards of accidental electrical contacts by all persons on site must be fully assessed. Generally apparatus operating at the higher voltages (such as 33 kV and above) is more expensive to repair if damaged than apparatus operating at lower voltages. Damage to underground cables can, in certain circumstances, cause widespread loss of electricity supplies for a long period. Therefore the electricity undertaker will usually require that greater care is shown and more protective works are undertaken where higher voltage cables are involved.

B4.5 Special Requirements

B4.5.1 Overhead Lines and Associated Apparatus

Many of the items listed under cables above also apply to overhead lines. However, special consideration needs to be given to the following -

i. The minimum height above ground of overhead lines, wires and cables are given in the Electricity Supply Regulations 1988 Part IV. An extract from Schedule 2 is given in the following table and refers to the minimum height of any point where the line is over a road accessible to vehicular traffic.

Not exceeding 33000 volts	5.8 m
Exceeding 33000 volts but not exceeding 66000 volts	6 m
Exceeding 66000 volts but not exceeding 132000 volts	6.7 m
Exceeding 132000 volts but not exceeding 275000 volts	7 m
Exceeding 275000 volts but not exceeding 440000 volts	7.3 m

Where not accessible to vehicular traffic, slightly reduced clearances are permitted in the Regulations.
On roads where abnormally high loads are likely a minimum clearance of 10 m is typical.

ii. Where civil engineering work, such as the removal of spoil, takes place near to towers and other overhead lines, their foundations can be weakened, thereby causing potential instability to the structure. In these cases there has to be close liaison between the undertaker and highway authority*.

iii. Where road widening schemes cause existing towers, poles and stays to be placed in such a position that they can subsequently cause danger or inconvenience to the public, they must be repositioned or removed. The undertaker's works may also require work on adjoining land. Where easements are necessary the scheme for the diversion of any works may have to be increased to satisfy landowners' wishes.

iv. Maintenance of appropriate clearances to ground and other obstacles can result in the need to replace or resite existing structures with taller and possibly wider base structures. Thus if an embankment is built under an overhead line, taller poles or towers may be necessary in order to maintain sufficient clearances. Associated structures may also be affected.

B4.5.2 Higher Voltage Underground Apparatus

Apparatus operating at the highest voltages of 132, 275 and 400 kV and associated equipment such as that indicated in section A4.4 usually requires special consideration because of the very high cost of diverting it and of the serious effect on electricity supplies if the apparatus develops faults in service. Undertakers would be unwilling to move such cables, whatever the type of road involved, because, being of a highly stressed, pressurised design (ie. oil-filled or gas pressure) they are very likely to be weakened electrically or suffer mechanical damage if disturbed. If it were absolutely necessary to move such high voltage circuits, it would be preferable to do a complete diversion involving the installation of new cables. Furthermore, where roads are improved or reconstructed, the depth of cables should not be unduly increased because increased cover adversely affects the current rating of cables. In other words, it would not be possible to transmit the required energy if the cables were laid at greater depths than originally allowed for at the design stage.

The fault incidence on the 132, 275 and 400 kV transmission apparatus is usually low, therefore it is unlikely that the repair of faults, assuming the cables are not moved in a road which is to be widened, would cause frequent inconvenience to traffic. Most of these circuits are installed beneath carriageways and have certain ancillary apparatus associated with them to which access is required about every six months, e.g. link boxes connected to the joints of cross-bonded cable systems, oil-pressure tanks provided at about one mile intervals for feeding oil and controlling the pressure of oil-filled cables. Such apparatus is usually installed in brick pits or buried concrete tubes either adjacent to the cable route or in the verge or footway alongside the cables.

Where existing roads are widened, therefore, it may be desirable, in certain instances, for ancillary apparatus to be moved to ensure that convenient access to it can be obtained.

APPENDIX C

PROCEDURES FOR NECESSARY MEASURES IN RELATION TO UNDERTAKERS' APPARATUS

C1 INTRODUCTION

In this appendix the term 'relevant authority' encompasses highway*, bridge or transport authority as appropriate.

C1.1 General

This Appendix sets out the recommended procedures to be adopted for consultation, planning and the execution of any undertakers' works that may be required as the result of the relevant authority's works. The aim is to avoid delays in the programming of the works and claims from the contractor, to safeguard undertakers' apparatus from damage and to minimise disruption to traffic flow.

C1.2 Outline of Procedure

While the procedures set out in this Appendix represent those which would normally be followed for larger-scale works, they may, depending on local circumstances and agreement, be foreshortened by the omission of certain of the following stages.

Seven basic stages are involved as follows -

- i. preliminary inquiries
- ii. draft scheme and budget estimates
- iii. detailed scheme and detailed estimates
- iv. formal notice and advance orders
- v. selection of contractor and issue of main orders
- vi. construction
- vii. financial monitoring and payment.

The foregoing procedures are summarised in Table 1 (see page 73).

For smaller-scale works, for example, the preliminary enquiry and draft scheme stages could be omitted and the process could commence with the submission of a detailed scheme to the undertakers by the authority. This would, in particular, apply in the case of smaller schemes or other schemes which may have minimal effect on undertakers' apparatus.

C1.3 Responses to Enquiries

The proposed response times in sections C2, C3 and C4 may have to be extended, by mutual agreement. This would be necessary when for example, it is necessary to make trial holes, when specialist suppliers have to be approached, or when notices have to be served before on-site investigations can be carried out.

C1.4 Establishing the Position of Undertaker's Apparatus

The existence of undertaker's apparatus is likely to have an important bearing on the cost of a scheme. This section gives guidance on information on the position of such apparatus required at the various stages of a scheme. It must be recognised that the records of the position of some older plant may not be satisfactory for the purposes of planning diversionary works.

At the preliminary inquiry stage, undertakers should provide information on their apparatus from what is available from their records and draw attention to any limitations in the quality of this information.

At the draft scheme stage, if the undertaker is not confident of the general position and nature of the apparatus, it should take any necessary steps to determine this information, this would be at its expense. Note that trial holing works in the street would need submission of the appropriate notices (see the Code of Practice for the Co-ordination of Street Works and Works for Road Purposes and Related Matters).

Trial holing can only be regarded as a sampling method and confirmation of the position of apparatus between trial holes (or other confirmed locations, such as manholes) can, for some apparatus, only be obtained by exposing the whole run of the apparatus. It cannot always be assumed that apparatus runs in straight lines and deviations may have been necessary at the time of installation to avoid obstructions or moved with or without the undertaker's knowledge subsquently. Careful exposure of the apparatus would normally be regarded as taking place as part of the works undertaken by the authority or their contractor at the construction stage.

If extra trial holing is requested by the authority before issuing of the formal order, it would be appropriate for the costs to be borne wholly or partly by the authority.

C2 PRELIMINARY INQUIRIES

At this stage the highway authority* would seek from the undertakers details of their apparatus within a specific section of the scheme which is being considered for improvement without making any commitment to the scheme. Undertakers should provide such information as they have available from records and draw attention to any likely special problems which could arise from the authority's works (see section C1.4). Information should be provided normally within 10 working days.

C3 DRAFT SCHEMES AND BUDGET ESTIMATES

The authority should submit details of the proposed scheme to undertakers. They will respond with preliminary details of the effects on their apparatus together with budget estimates for the necessary works and an indication of any special requirements involved such as -

    i.   items of equipment on long delivery and the need for advance ordering

    ii.   interruption of supplies to consumers

    iii.   disconnection of supplies to premises which are to be demolished

    iv.   special wayleave agreements associated with the diversion of their apparatus

    v.   early access to sites for the construction of special structures such as sub-stations, pressure regulation stations etc.

    vi.   planning consents or special ministerial consents.

Budget estimates provided by undertakers should include all costs likely to arise from the necessary measures in consequence of the authority's works as far as can reasonably be assessed at the draft design stage, including administration and supervision charges, and specifying a base date. These estimates should be provided normally within 20 working days.

Where undertakers are not aware of the general position in line and depth of their apparatus, they should at this stage take any necessary steps to determine this information (see section C1.4).

This stage may be followed by discussions between the authority and undertakers, either separately or jointly, to consider any modifications to the scheme which may assist in facilitating the programming of the works and/or reducing the cost of diversionary works.

C4 THE FINAL DETAILED SCHEME AND DETAILED ESTIMATES

Following joint discussions, the authority should submit to each undertaker details of the final design with working drawings and an outline programme. The undertakers should respond, normally within 25 working days, by providing details of their requirements (if there is a requirement to provide more than one detailed estimate, the utility may charge for such additional estimates) as follows -

i. Description of the necessary measures, clearly stating the reason for diversion or protection and giving details of the existing apparatus affected, such as lengths and sizes of pipes/cables/ducts, depths of cover and ages. See section C1.4 on the possible need to confirm positions of apparatus.

ii. A detailed specification of the works required as appropriate -

   (a) details of all replacement apparatus, i.e. lengths, type, material, size and routes, drawing attention to those materials with long-lead times on delivery.

   (b) details of all protection work

   (c) advance or off-site works

   (d) method of construction and sequence of operations where these have a significant effect on cost or programme

   (e) arrangements for delivery of materials and storage requirements

   (f) route and level requirements, trench dimensions, methods of excavation and assumed ground conditions

   (g) reinstatement requirements, backfill specification, type of reinstatement (temporary, interim, permanent) and removal of surplus spoil

   (h) details of temporary works

   (i) any special requirements, e.g. provision of temporary accommodation for staff

   (j) details of the requirements for commissioning of apparatus, e.g. original apparatus having to remain commissioned until all services are transferred

   (k) method of dealing with apparatus made redundant by the scheme, e.g. recovered or abandoned in situ.

iii. A detailed estimate with itemised costs also to include -

   (a) overhead and supervision charges

   (b) details of the likely allowance for deferment of renewal based on the formula set out in Appendix E and for betterment given in Appendix F.

   (c) allowance for any materials recovered

   (d) the base date for the estimate its period of validity, and method of dealing with inflation; for example, Baxter indexation.

   NOTE (If works are to be phased over a long period of time, estimates for each phase may be appropriate)

iv. Provisional programmes and timescales for works including as appropriate:

   (a) site works

(b) off-site works

(c) time for obtaining materials with a long lead-time on delivery

(d) land purchase

(e) wayleaves acquisition

v. It may be advantageous for the civil engineering element of the undertaker's works to be undertaken by the authority's contractor as part of the main works. Such an approach can yield benefits in reduction of cost and simplification of programme co-ordination. If the option is pursued it will be necessary for the undertaker to provide all the necessary information for the civil engineering work for inclusion in the authority's tender documents. If this approach is adopted then the undertaker is likely to require inspection of the works during construction. Civil engineering works costs will have to be separately identified since they remain part of the total undertakers' costs - see Section C9.9(i).

This stage may be followed by further discussions between the authority and undertaker to consider modifications to the works in order to minimise costs.

C5 FORMAL NOTICE AND ADVANCE ORDERS

Formal notice of the authority's intention to proceed with the scheme should be served on the undertakers. Unless otherwise stated, the formal notice will be taken as an instruction to the undertaker to proceed with advance ordering of those materials which have long delivery periods and to undertake those works which require more extensive preparation. It is possible for the authority to serve formal notice at an earlier stage of the procedure but it is preferable to have done so following consultation and receipt of proposals from the undertakers.

The undertakers should acknowledge receipt of the notice and respond with a detailed specification, itemised estimate and programme if not already submitted.

C6 SELECTION OF CONTRACTOR AND ISSUE OF MAIN ORDERS

The authority will incorporate the undertakers' proposals and programmes into the tender documents. A contractor is in due course selected and the authority will appoint an Engineer. The main orders from the authority for the necessary measures will be issued to the undertakers.

The undertakers should be advised of the name of the contractor and his site staff, the name of the Engineer and his representatives, and the commencing date of the contract. Similarly, the authority and the Engineer should be notified by the undertakers of the names of their respective representatives and, if appropriate, their contractor.

In the case of highway works*, the authority will normally let a contract within three weeks (but longer for major contracts) from the closing date for the receipt of tenders. The contractor will be expected to commence works as soon as is reasonably possible thereafter.

C7 CONSTRUCTION STAGE

C7.1 Programme and Coordination

Joint discussions should then take place between the Engineer, the contractor's representative, and the undertakers to establish detailed programming, methods of working and the general coordination of the works. It is anticipated that these arrangements will follow the details previously given by the undertakers. However, variations may be agreed between the contractor and the undertakers, and approved by the authority, including any cost changes. Once agreement has been reached on these matters, any subsequent variation could give rise to a claim from either party (see Section C8).

Regular meetings should take place between the Engineer, the contractor and the undertaker or their representatives during the progress of the contract to ensure as far as practicable its smooth operation.

   i.   The Engineer should ensure that -

   (a) Undertakers are immediately advised of programme or layout changes. These may, for example, drastically affect undertakers' work programmes as it may not be possible for undertakers' apparatus to be taken out of service due to operational restrictions and/or without due notice to existing consumers.

   (b) As far as practicable the undertakers' routes are identified, checked and cleared of all obstructions to enable their works to be carried out.

   (c) A master plan of all the undertakers' apparatus is prepared and maintained on site from information provided by undertakers. It should always be assumed that all apparatus is 'in service' unless the undertaker responsible confirms otherwise (see also Section C1.4).

   (d) Nothing is done to prejudice the safety, access to or operation of any undertakers' apparatus.

   (e) Any damage caused to undertakers' apparatus is immediately reported to the undertaker concerned.

   (f) Any necessary precautions are taken to protect undertakers' apparatus from damage by excessive loading or otherwise during the progress of the contract.

   ii. The undertakers should ensure that -

   (a) The Engineer is notified as soon as possible of any changes in the undertaker's programme of work.

   (b) The authority is advised of any significant cost variations arising from programme changes, with appropriate details.

   (c) Nothing is done on site which would prejudice the safety or operation of the works or other undertakers' apparatus.

   (d) Excavations are properly backfilled in accordance with the specification and surplus spoil disposed of.

   (e) Restoration, where specified, is carried out immediately following completion of their work.

   (f) The Engineer is provided as promptly as possible with details of work done for recording on the master plan.

C7.2 Changes During Construction

It is important that any variation, error or change of programming in the implementation of the scheme or the undertakers' works be notified to the affected parties or Engineer as soon as possible and confirmed in writing forthwith.

C8   CLAIMS

If the authority or the undertaker fails to comply with an agreement between them as to the necessary measures, by virtue of section 84 of the Act (for England and Wales) or section 143 of the Act (for Scotland) the authority or undertaker is liable to compensate the other in respect of any loss or damage resulting from the failure.

Claims may arise from several sources, for example -

   i.   from the authority when their contractor suffers delays to the progress of the contract because an undertaker has been unable to meet the agreed programme. Such a claim may cover for standing time for plant, machinery and labour, together with associated overheads. Extended delays may give rise to requests for an extension to the contract period and associated establishment charges for the additional time involved.

ii. from an undertaker to the authority where, because of a delay in the authority's or contractor's programme, the undertaker has been unable to execute his programmed works. Such a claim may cover for standing time for plant, machinery and labour and other associated costs.

Every effort should be made to avoid or mitigate claims by early consultations between appropriate parties when delays are likely to occur to programmes. All claims should be notified as soon as possible and then supported in writing, setting out full details and costs and justification for the claim. Claims should only be entertained when it can be clearly shown that a financial loss has occurred as the direct result of the non-performance of another party. Claims will not normally be entertained when conditions of 'force majeure' prevail.

Claims may also arise, directly or indirectly, from delays suffered by one undertaker as the result of the default of another. Undertakers should endeavour to resolve these claims in the first place without reference to the authority or the authority's contractor.

C9 INVOICING, PAYMENT AND FINANCIAL MONITORING

C 9.1 Introduction

This section describes the procedures to be adopted to satisfy the financial principles given in section 9 and laid down in Regulations. It provides for the following situations:

i. invoicing and payment for undertaker's works by a standard invoice (see section C9.2), rendered and paid in advance followed on completion of the works by a final invoice.

ii. invoicing and payment for works of long duration where payments by instalments is agreed.

iii. arrangements when invoices are not settled within the specified time periods.

iv. arrangements for special circumstances.

C9.2 Construction of the Standard Invoice

Invoices will be presented by undertakers in a form which shows how they have been constructed and include the following information:

A = total estimated cost of works (or stage of works), itemised to reflect the detailed estimate provided under section C4(iii)

B = 82% of A, being the proportion chargeable to the authority where advance payment is agreed beforehand, or 100% where not agreed

C = 75% of B, being the proportion of B which is required to be paid in advance of work commencing

D = 25% of B, or the outstanding amount on finalisation of actual accounts for the works (or stage of works) to be paid within 30 days of the presentation of and agreement to the final account.

The invoice(s) for advance payment will be for amount C.

The final invoice after completion of works (or stage of works) will be for amount D.

Where payment by instalments has been agreed (see section C9.3.ii. below), the level of detail given in interim invoices may be reduced but the final account must be fully detailed.

C9.3 Advance Payments

(Item C of Section C9.2)

i. **Case 1 - Single Payment**

This case will typically apply to works of short duration where the benefits of payment by instalments are outweighed by administration costs. It is recommended that works of less than 3 months' duration should be prepaid by the standard invoice (see section C9.2), subject to final invoicing.

ii. **Case 2 - Payment in Instalments**

Where works extend over a long period, payment may, by mutual agreement, be by instalments. There are two possibilities:

(a) the total costs are divided by the estimated duration of the works in months and invoices submitted for equal monthly payments in advance and also settled on a monthly basis.

(b) where the works are of long duration but where the costs incurred are liable to fluctuate greatly over the period, then agreement should be reached as to the timing of invoices and the amounts to be invoiced over the period.

Where invoices are to be monthly or at some other agreed interval then, at the choice of the authority, the undertakers can either

- invoice the authority each month for the agreed amount, or

- raise one invoice for the total amount, showing the instalments agreed and the dates when due.

In response to current VAT Regulations, when the latter option is chosen, because VAT will become due when the invoice is rendered, the first payment must be increased by the total amount of VAT due, ie. the VAT element will not be spread over the instalments.

The requirements for timely settlement or any invoice for advance payment are detailed in section 9.5 below.

C9.4 Alternative to Advance Payment

Where an authority does not wish to pay in advance of the undertaker commencing works, a single invoice will be issued on completion of the works or agreed stages of works. The 18% allowance will not then be deducted by the undertaker from the cost of the works. Payment of interim invoices should be made within 30 days of presentation and payment of the final invoice within 30 days of agreement.

C9.5 Late Payment and Late Commencement

i. Regulations require the undertaker's invoice for advance payment to be settled within 35 days. If payment is late then interest may be charged on the sum due on a daily basis for the period between 36 days from the invoice date until payment is received.

ii. Where the undertaker is late in starting his works after the advance invoice has been paid other than by agreement with the authority, then the authority may be reimbursed at a rate equivalent to interest for the period of delay.

See section 9.8 for further detail on interest payments.

C9.6 Final Payment

The final payment should be made on the basis of an itemised invoice detailing the works which are rechargeable. The invoice should be presented in a similar format to the detailed estimate described in section C4. The invoice should include details of all agreed variations from the original estimate.

The undertaker will calculate his final costs incurred on the scheme including allowances for deferment of renewal as defined in Appendix E, betterment as defined to Appendix F, and credit for recovered or scrap materials. The 18% allowance will apply to the resulting total.

C9.7 Settlement of Final Invoice

Payment in final settlement of the total charge for the works shall be made within 30 days of the works being invoiced and agreed. If the undertaker is not notified of any disagreement then interest will be payable after 30 days.

C9.8 Interest or Reimbursement Payments

Where interest is required to be paid or reimbursement made, this will be at one percent above the prevailing base rate from the due date until the actual payment of the sum bearing interest or requiring reimbursement.

It is recommended that such payments should only be required when the sums involved exceed £100 or where there is evidence of persistent delays either in settling invoices or late start of work for which the undertaker has been paid.

C9.9 Special Cases

    i. **Works carried out by the authority on behalf of the undertaker.** When it has been agreed that the authority will carry out duct work or other civil engineering work on behalf of the undertaker, then the authority will provide a full estimate of the works to be carried out as detailed in section C4 (v) of this appendix.

    The estimate for such works will be deducted from cost B in section C9.2 above. Costs C and D will then be calculated on the revised value of B. When the final calculation is done, the net effect should be the same as if the authority had been given an allowance of 18% on the total costs involved, including any civil engineering work carried out directly by it.

    ii. **Advance ordering of special materials.** Where special materials have to be ordered in advance for the works (see section C5) the authority may be billed when these have been received by the undertaker. The standard rules for allowances given in section C9.2 will apply and invoices constructed accordingly. Subsequent invoices should take account of such advance payments.

    Should works be cancelled after such special materials have been ordered, the authority would be liable for the full cost of such materials if already delivered or cancellation charges if not delivered. The cost to the authority should allow for the value of any materials put to alternative use.

    iii. **Works not funded by the authority.** The element of cost which will be considered for the application of the 18% allowance will be for work which is undertaken solely for the authority. Any contribution made to the authority's scheme by a developer or other body either in cash or kind will not be eligible for the allowance. The authority will be expected to supply cost apportionment information when the formal order for the work is placed for schemes which involve a commercial development content so that the appropriate allowance can be calculated.

    Where the development content is a substantial proportion of the total works, the undertaker is likely to require full advance payment equivalent to the developer's contribution unless the authority is prepared to indemnify the undertaker against any loss should the developer's work not proceed.

C9.10 Monitoring

The undertaker should provide the authority with monthly reports on costs incurred and projected outturn. Where the actual costs vary significantly from the estimated (either in programme or cost), revised monthly payments may be agreed.

Reasonable facilities should be made available to permit either party to verify that the charges being made by the other party are allowable and reasonable.

TABLE 1

OUTLINE OF PROCEDURES IN PLANNING AND IMPLEMENTATION OF MAJOR SCHEMES

STAGE	ACTION OF THE AUTHORITY	UNDERTAKER'S ACTION	REF
Preliminary Enquiries	request location of undertakers' apparatus		C2
		provide information, normally within 10 working days	
Draft Scheme(s) and Budget Estimates	prepare draft scheme including any special aspects		C3
		provide budget estimate and indicate special requirements normally within 20 working days	
	amend schemes to reduce costs		
		assist in studying alternatives	
The Final Detailed Scheme and Detailed Estimates	prepared detailed scheme and ask undertakers for detailed estimate and specification		C4
		provide detailed estimate and specification, normally within 25 working days	
	(consider further amendments to the scheme)		
Formal Notice and Advance Orders	place order with undertakers for long lead-time items		C5
		order long lead-time items	
	serve notice upon undertakers		
		respond to notice	
Selection of Contractor and Issue Main Orders	- place order on undertakers - appoint contractor and advise him of undertakers' requirements - appoint engineer		C6
		notify the authority and contarctor of representatives	
Construction	arrange discussions and liaison		C7
		confirm timescales and liaise with the Engineer and contractors	
Financial Monitoring and Payment		present invoices (interim if appropriate)	C9
	settle invoices		

APPENDIX D

STOPPING UP AND DIVERSION ORDERS

This Appendix briefly summarises the main powers under which stopping up and diversion orders may be made and the protection provided for the rights of undertakers in such cases. It does not purport to set out the full legal position.

ENGLAND AND WALES

Type of Highway	Powers	Order Making Authority
1. Roads affected by the construction/improvement of a trunk road (or special road provided by the Secretary of State)	Highways Act 1980, sections 14 and 18	Secretary of State
2. Roads affected by the construction/improvement of a classified road (or special road provided by the local authority)	Highways Act 1980, sections 14 and 18	Local highway authority (Confirmation by Secretary of State)
3. Roads (other than trunk or special roads) no longer necessary (or diverted to become more commodious to the public)	Highways Act 1980, section 116	Magistrate's Court
4. Highways affected by development	Town and Country Planning Act 1990, section 247	Secretary of State
5. Highways crossing or entering the route of a proposed new highway	Town and Country Planning Act 1990, section 248	Secretary of State

UNDERTAKERS OTHER THAN TELECOMMUNICATIONS OPERATORS

Under sections 14 and 18, Highways Act 1980

Undertakers' rights to retain apparatus in the old highway may be extinguished unless provision is specifically included in the order to protect such rights. Sections 21 and 22 apply the code set out in sections 271 to 282 of the Town and Country Planning Act 1990 which enables the highway authority to require the undertaker to move his apparatus, but obliges them to pay the undertaker's costs (section 279). The Town and Country Planning Act code also allows the undertaker to require that his apparatus be moved, again at the expense of the highway authority; provision exists for the highway authority to serve a counter-notice.

Under section 116 (and Schedule 12), Highways Act 1980

Undertakers retain the same powers and rights in respect of their apparatus in the highway as though the stopping up order had not been made. Schedule 12, Part III, paragraph 5, of the 1980 Act provides for the apparatus to be moved at the option of the undertaker or at the reasonable request of the highway authority and the highway authority must pay the cost of essential works.

Under sections 247 and 248, Town and Country Planning Act 1990

Undertakers lose their powers to retain apparatus in the old highway unless provision is made under section 247(4)(b) to preserve those rights. The Secretary of State can make such provision as appears to him to be necessary or expedient under section 247(2). If undertakers' rights are extinguished under the code, the relevant authority must pay the undertakers' costs.

TELECOMMUNICATIONS OPERATORS

Under sections 14 and 18, Highways Act 1980

The protection provided by section 334 of the Highways Act 1980 as amended applies. In essence the operator retains the same power in respect of the apparatus as if the order under sections 14 and 18 had not been made.

Under section 116, Highways Act 1980

The protection provided in section 334 of the Highways Act 1980 as amended by the Telecommunications Act 1984, Schedule 4, paragraph 76(4), applies. In essence the highway authority must serve notice of the granting of the stopping up order on the operator. The operator has 3 months to serve a counter-notice requiring the highway authority to pay for the removal of apparatus affected by the order.

Under sections 247, 248 and 249, Town and Country Planning Act 1990

The protection provided by section 256 of the Town and Country Planning Act 1990 applies. In essence, the operator retains the same powers in respect of the apparatus as if the order under section 248 had not been made.

SCOTLAND

Type of Roads	Powers	Order Making Authority
1. Roads affected by the construction/improvement of special roads	Roads (Scotland) Act 1984, section 9	Secretary of State for Scotland, or local roads authority confirmed by Secretary of State
2. Roads affected by the construction/improvement of public roads other than special roads under section 9(1)(c) of the Roads (Scotland) Act 1984	Roads (Scotland) Act 1984, section 12	Secretary of State for Scotland, or local roads authority (confirmed by Secretary of State where objections are not withdrawn)
3. Roads (other than those where section 12 or section 9(1)(c) of the Roads (Scotland) Act 1984 would apply) which have become dangerous, or unnecessary	Roads (Scotland) Act 1984, section 68	Secretary of State for Scotland, or local roads authority (confirmed by Secretary of State where objections are not withdrawn)
4. Roads affected by development	Town and Country Planning (Scotland) Act 1972, section 198	Secretary of State for Scotland

5. Roads other than trunk or special roads affected by development	Town and Country Planning (Scotland) Act 1972, section 198A and section 206	Local Planning Authority confirmed by Secretary of State for Scotland in accordance with section 206 where the order is opposed
6. Footpaths and bridleways affected by development	Town and Country Planning (Scotland) Act 1972, section 199	Local Planning Authority (confirmed by Secretary of State for Scotland where the order is opposed)

UNDERTAKERS OTHER THAN TELECOMMUNICATIONS OPERATORS

Under section 9 of the Roads (Scotland) Act 1984

A roads authority may remove an undertaker's apparatus from land acquired or appropriated by them in pursuance of a special road scheme. Section 134 of the Roads (Scotland) Act applies the provisions of sections 219 and 220 and 222 - 225 of the Town and Country Planning (Scotland) Act 1972. This sets out the procedures for a roads authority to serve notice of the extinguishment of the right to place and renew apparatus, and the procedure for a statutory undertaker to serve a counter-notice.

Under sections 12 and 68 of the Roads (Scotland) Act 1984

Orders under these sections must include provision for the preservation of statutory undertakers' rights in respect of apparatus in the road at the time the order is made.

Under sections 198, 198A and 199, Town and Country Planning (Scotland) Act 1972

Orders made under these sections may include provision for the preservation of any rights of statutory undertakers in respect of apparatus in the road at the time the order is made.

TELECOMMUNICATIONS OPERATORS

Under sections 9, 12 and 68 of the Roads (Scotland) Act 1984

The protection provided by section 132 of the Roads (Scotland) Act 1984 applies. In essence the operator retains the same power in respect of the apparatus as if the stopping up order had not come into force, but any person entitled to land over which the road subsisted might require the alteration of the apparatus.

Under sections 198, 198A and 199 of the Town and Country Planning (Scotland) Act 1972

Section 209 (provision as to telegraphic lines) of the Town and Country Planning (Scotland) Act 1972 is amended by Schedule 4, paragraph 54, of the Telecommunications Act 1984. Essentially for orders under sections 198 and 198A the operator retains the same power in respect of any telecommunications apparatus as if the stopping up order had not become operative, but any person entitled to land over which the road subsists, and any local roads authority is entitled to require the alteration of the apparatus. For orders under section 199 the planning authority must give notice of the making of the order. Thereafter the operator must remove or give notice of their intention of removing apparatus within 3 months of the date of the order stopping up the bridleway or footway, or else the apparatus will be deemed to have been abandoned. The operator is entitled to recover certain expenses from the planning authority.

APPENDIX E

DEFERMENT OF THE TIME FOR RENEWAL

E1 Deferment of the Time for Renewal

E1.1 Any financial benefit conferred on undertakers by reason of the deferment of the time for renewal of the apparatus in the ordinary course shall reduce the payment from the relevant authority and will be calculated from actual costs by the "Bacon and Woodrow" formula. The following guidance applies -

(a) An allowance is due only if apparatus is more than 7½ years old.

(b) An allowance is due only on individual lengths of diversion of greater than 100 metres. For optical fibre telecommunications cables, the lengths must be 500 metres or greater. For electricity cables the lengths have to be more than 500 metres for circuits of voltages of 33 kV and above; 250 metres for all auxiliary cables and circuits of voltages above medium voltage but less than 33 kV, and 100 metres for medium voltage circuits.

(c) The costs of disconnecting and reconnecting into existing apparatus should be excluded from the costs used to calculate the allowance.

(d) Any allowance for betterment should be excluded from the costs used to calculate the allowance.

(e) Total costs should be used in the calculations.

(f) The full value of any recovered material (including scrap) should be deducted from the payment made by the relevant authority.

E1.2 The interest rate is set at 6% per annum but may be reviewed at the request of either side but not more frequently than three years. The interest rate is intended to represent a long run "real" rate, ie. interest less inflation.

E1.3 The estimated full life of apparatus should be as set out in section E3.

E2 The Bacon and Woodrow Formula

For the calculation of financial benefit from the deferment of the time of renewal, the Bacon and Woodrow formula should be used as follows -

$$B = C \left[\frac{(1 + R)^b - 1}{(1 + R)^L} \right]$$

where
C = Cost of undertakers' works
R = Rate of interest
L = Number of years of estimated full life of apparatus
b = Number of years of expired life of apparatus
B = Financial benefit

Table 2 is a ready reckoner which tabulates the value of the factor B for different full and expired lives and uses an interest rate of 6% per annum.

E3 Life of Apparatus

As the basis for application of the formula, the following have been accepted as the normal lives of apparatus with nil residual value.

Electricity -

    all types of apparatus ....80 years

Telecommunications -

    all types of apparatus ....60 years

Gas -

Cast iron (including ductile and spun iron) and protected steel pipes:

    Up to 8" (203.2 mm) OD .... 80 years
    8" (203.2 mm) to less than 12" (304.8 mm) OD ....100 years
    12" (304.8 mm) OD and over .... by agreement
    Polyethylene pipes ....120 years

Water -

Cast iron (including ductile and spun iron) pipes:

    Up to 8" (203.2 mm) nominal bore... 80 years
    8" (203.2 mm) to less than 12" (304.8 mm)
    nominal bore ....100 years
    12" (304.8 mm) nominal bore and over .... by agreement

Protected steel pipes -

    Since few protected steel pipes of less than 12"
    304.8 mm) nominal bore are now used in the
    Water Industry ....by agreement

Asbestos cement pipes -

    Up to 8" (203.2 mm) nominal bore...100 years
    8" (203.2 mm) to less than 12" (304.8 mm)
    nominal bore ....120 years
    12" (304.8 mm) nominal bore and over ....by agreement

Polyethylene pipes ....**120 years**

PVC pipes ....**120 years**

Sewer pipes ....**by agreement**

TABLE 2

TABLE APPLICABLE TO 6 PER CENT PER ANNUM INTEREST

To use this table, select from the first column the expired life (to the nearest year) of the apparatus, and against this in the relevant full life column read the factor by which the cost of undertakers' works needs to be multiplied to give the appropriate amount of financial benefit.

Example of Calculation of Allowance for Deferment of Renewal

Assume that the apparatus has, because of diversion, only achieved 25 years of its normal expected life of 80 years. The replacement apparatus, being new, will also have an expected life of 80 years from the time of installation. Benefit will therefore accrue to the undertaker in that apparatus in that location will not now have to be replaced in 55 years' time but 80 years hence. Allowance is then made for this benefit as follows:

assume new apparatus costs £100,000

estimated full life	=	80 years
expired part of life	=	25 years
from Table 2, factor	=	0.03112

amount therefore to be used when calculating cost to authority

=	100000 - (100000 x 0.03112)
=	£96,888

Estimated Full Life of Apparatus

Expired Life of Apparatus (Years)	60 years	80 years	100 years	120 years
8	.01800	.00561	.00175	.00055
9	.02090	.00652	.00203	.00063
10	.02397	.00748	.00233	.00073
11	.02723	.00849	.00265	.00083
12	.03068	.00957	.00298	.00093
13	.03434	.01071	.00334	.00104
14	.03822	.01192	.00372	.00116
15	.04234	.01320	.00412	.00129
16	.04670	.01456	.00454	.00142
17	.05132	.01600	.00499	.00155
18	.05621	.01753	.00547	.00171
19	.06141	.01915	.00597	.00186
20	.06691	.02086	.00651	.00203
21	.07274	.02268	.00707	.00220
22	.07893	.02461	.00767	.00239
23	.08548	.02665	.00831	.00259
24	.09243	.02882	.00899	.00280
25	.09979	.03112	.00970	.00302

Expired Life of Apparatus (Years)	60 years	80 years	100 years	120 years
26	.10760	.03355	.01046	.00326
27	.11587	.03613	.01127	.00351
28	.12464	.03886	.01212	.00378
29	.13394	.04176	.01302	.00406
30	.14380	.04484	.01398	.00436
31	.15424	.04810	.01500	.00467
32	.16532	.05155	.01607	.00501
33	.17705	.05521	.01721	.00537
34	.18950	.05909	.01842	.00574
35	.20269	.06320	.01971	.00614
36	.21667	.06756	.02107	.00657
37	.23148	.07218	.02251	.00702
38	.24719	.07708	.02403	.00749
39	.26384	.08227	.02565	.00800
40	.28149	.08777	.02737	.00853
41	.30020	.09360	.02919	.00910
42	.32003	.09979	.03111	.00970
43	.34105	.10364	.03316	.01034
44	.36333	.11329	.03532	.01101
45	.38695	.12065	.03762	.01173
46	.41199	.12846	.04005	.01249
47	.43853	.13673	.04263	.01329
48	.46666	.14551	.04537	.01415
49	.49647	.15480	.04827	.01505
50	.52808	.16466	.05134	.01601
51	.56158	.17511	.05460	.01702
52	.59710	.18618	.05805	.01810
53	.63474	.19792	.06171	.01924
54	.67465	.21036	.06559	.02045
55	.71694	.22355	.06970	.02173
56	.76178	.23753	.07406	.02309
57	.80931	.25235	.07868	.02453
58	.85968	.26805	.08358	.02606
59	.91308	.28470	.08877	.02768
60	.96970	.30235	.09428	.02939
61		.32106	.10011	.03121
62		.34089	.10629	.03314
63		.36191	.11285	.03518
64		.38419	.11979	.03735
65		.40781	.12716	.03965
66		.43285	.13497	.04208
67		.45939	.14324	.04466
68		.48752	.15201	.04740
69		.51734	.16131	.05030
70		.54894	.17116	.05337
71		.58245	.18161	.05663
72		.61796	.19268	.06008
73		.65561	.20442	.06374
74		.69551	.21686	.06762
75		.73781	.23005	.07173
76		.78264	.24403	.07609
77		.83017	.25885	.08071
78		.88054	.27456	.08561
79		.93394	.29121	.09080

Expired Life of Apparatus (Years)	60 years	80 years	100 years	120 years
80		.99056	.30886	.09630
81			.32757	.10214
82			.34740	.10832
83			.36842	.11487
84			.39070	.12182
85			.41432	.12919
86			.43935	.13699
87			.46589	.14527
88			.49402	.15404
89			.52385	.16333
90			.55545	.17319
91			.58895	.18364
92			.62447	.19471
93			.66211	.20645
94			.70201	.21889
95			.74431	.23208
96			.78915	.24606
97			.83667	.26088
98			.88705	.27659
99			.94045	.29324
100			.99706	.31088
101				.32959
102				.34942
103				.37044
104				.39273
105				.41635
106				.44138
107				.46792
108				.49605
109				.52587
110				.55747
111				.59098
112				.62649
113				.66414
114				.70404
115				.74634
116				.79117
117				.83870
118				.88908
119				.94248
120				.99908

APPENDIX F

BETTERMENT

F1   Conditions for Allowance

A financial benefit should be allowed for betterment which includes -

(a)   Increasing the capacity of the apparatus, except where this has been solely due to using the nearest currently available size.

(b)   Using a material or type of material which enhances the capability of the network. This does not include using the material or type of apparatus which is in present day use for the same duty, e.g. polyethylene would be used for diverting small diameter cast iron low pressure gas mains.

The allowance to be given for increasing the capacity or enhancing the duty of the apparatus should be the difference between the cost of laying the increased capacity or enhanced duty apparatus and the estimated cost of laying the same capacity or same duty apparatus (or nearest equivalent).

F2   Presentation of Calculations

Details on what information should be shown in the calculations is covered in Appendix C.

GLOSSARY

(NOTE: References in this Glossary to numbered sections are to sections of the New Roads and Street Works Act 1991, unless otherwise stated.)

Act: unless otherwise stated, in this Code of Practice and Glossary, 'the Act' refers to the New Roads and Street Works Act 1991.

apparatus: includes any structure for the lodging therein of apparatus or for gaining access to apparatus (sections 105 and 164).

authority: means the same as 'relevant authority' (q.v.)

bridge: reference to a bridge includes so much of any street (or in Scotland any road) as gives access to the bridge, and any embankment, retaining wall or other work or substance supporting or protecting that part of the street (or road) (sections 88 and 147).

bridge authority: the authority, body or person in whom a bridge is vested. (sections 88 and 147).

carriageway : in England and Wales that part of the highway, other than a cycle track, set aside for the passage of vehicles (section 329 of the Highways Act 1980).

in Scotland that part of a road over which there is a public right of passage by vehicle, other than a right by pedal cycle only (Roads (Scotland) Act 1984, section 151(2)).

costs: the costs or expenses of taking any action shall be taken to include the relevant administrative expenses (of that authority, body or person) including general staff costs and overheads (sections 96 and 155).

engineer: as defined in the Institution of Civil Engineers Conditions of Contract.

footway: **Highways Act 1980** - (applicable in England and Wales)

a way comprised in a highway which also comprises a carriageway, being a way over which the public have a right of way on foot only.

Roads (Scotland) Act 1984 - section 151(2) (applicable in Scotland)

that part of a road associated with a carriageway over which there is a right of passage by foot.

footway and carriageway construction: all layers forming part of the road carriageway. A typical cross section is shown below for a road and the "construction" includes the wearing course, base course, road base and sub-base.

road construction:

```
          ┌─────────────────┐
          │ WEARING COURSE  │   Running Surface
          │   BASE COURSE   │
        { │    ROAD BASE    │
          │     SUB-BASE    │
          │    SUB-GRADE    │   Formation Level
          └─────────────────┘
```

highway: in England and Wales, includes the carriageway, verge and footway.

highway authority: for England and Wales in the case of trunk roads (which include most motorways), the Secretary of State for Transport or the Secretary of State for Wales; in the case of all other roads maintainable at the public expense, the County Council, Metropolitan Borough Council, London Borough Council, or Common Council of the City of London in whose area the road is situated. (The equivalent definition applicable in Scotland is given under 'roads authority'.)

Highway Authorities and Utilities Committee (HAUC): the national committee comprising representatives of the local authority associations and the National Joint Utilities Group which deals with matters of common interest.

maintainable highway: in England and Wales, a highway which for the purpose of the Highways Act 1980 is maintainable at the public expense. The equivalent definition applicable in Scotland is given under 'public road'.

major highway works: in England and Wales, works of any of the following descriptions executed by the highway authority in relation to a highway which consists of or includes a carriageway -

(a) reconstruction or widening of the highway,

(b) works carried out in exercise of the powers conferred by section 64 of the Highways Act 1980 (dual carriageways and roundabouts),

(c) substantial alteration of the level of the highway,

(d) provision, alteration of the position or width, or substantial alteration in the level of a carriageway, footway or cycle track in the highway,

(e) the construction or removal of a road hump within the meaning of section 90F of the Highways Act 1980,

(f) works carried out in exercise of the powers conferred by section 184 of the Highways Act 1980 (vehicle crossings over footways and verges),

(g) provision of a cattle-grid in the highway or works ancillary thereto, or

(h) tunnelling or boring under the highway
(section 86(3)).

In addition, section 86(4) states 'Works executed under section 184(9) of the Highways Act 1980 by a person other than a highway authority shall also be treated for the purposes of this Part as major highway works'.

The equivalent definition applicable in Scotland is given under 'major works for roads purposes'.

major bridge works: works for the replacement, reconstruction or substantial alteration of a bridge (sections 86 and 147).

major transport works: substantial works required for the purposes of a transport undertaking and executed in property held or used for the purposes of the undertaking (sections 91 and 150).

major works for roads purposes: in Scotland works of any of the following descriptions executed by the roads authority in relation to a road which consists of or includes a carriageway -

(a) reconstruction or widening of the road,

(b) substantial alteration of the level of the road,

(c) provision, alteration of the position or width, or substantial alteration in the level of a carriageway, footway or cycle track in the road,

(d) the construction or removal of a road hump within the meaning of section 40 of the Roads (Scotland) Act 1984,

(e) works carried out in exercise of the powers conferred by section 63 of the Roads (Scotland) Act 1984 (new access over verges and footways),

(f) provision of a cattle-grid in the road or works ancillary thereto,

(g) tunnelling or boring under the road (section 145(3)).

The equivalent definition applicable in England and Wales is given under 'major highway works'.

necessary measures: measures needing to be taken in relation to undertakers' apparatus in consequence of, or in order to facilitate the execution of major highway works or in Scotland major works for roads purposes, major bridge works, or major transport works (sections 84 and 143).

public road: in Scotland, includes the carriageway, verge and footway. Also, it is a road which a roads authority have a duty to maintain (New Roads and Street Works Act 1991, section 145). The equivalent definitions applicable for England and Wales are given under 'highway' and also 'maintainable highway'.

relevant authority: means, in relation to any works in a street or in Scotland a road, the street authority and the road works authority and also -

(a) where the works include the breaking up or opening of a public sewer in the street or road, the sewer authority,

(b) where the street or road is carried or crossed by a bridge vested in a transport authority, or crosses or is crossed by any other

	property held or used for the purposes of a transport authority, that authority, and,
	(c) where in any other case the street, or in Scotland a non-public road, is carried or crossed by a bridge, the bridge authority (sections 49 and 108).
road:	in Scotland, for the purpose of road works executed in terms of Part IV of the Act, means any way (other than a substitute road made under section 74(1) of the Roads (Scotland) Act 1984 or a waterway) whether or not there is over it a public right of passage and whether or not it is for the time being formed as a way; and the expression includes a square or court, or any part of a road.
	where a road passes over a bridge or through a tunnel, references to the road include that bridge or tunnel (section 107). The equivalent definition applicable in England and Wales is given under 'street'.
roads authority:	in Scotland, means in the case of trunk roads (which include most motorways), the Secretary of State for Scotland; in the case of all other public roads the Regional or Islands Councils.
road managers:	in Scotland means, in relation to a road which is not a 'public road', the authority, body or person liable to the public to maintain or repair the road or, if there is none, any authority, body or person having the management or control of the road (section 108). The equivalent definition applicable for England and Wales is given under 'street managers'.
road with special engineering difficulties:	see under definition of 'street with special engineering difficulties'.
road works:	see under definition of 'street works'.
road works authority:	in Scotland means -
	(a) if the road is a public road, the roads authority, and
	(b) if the road is not a public road, the road managers (section 108). The equivalent definition applicable in England and Wales is given under 'street authority'.
road works permission:	in Scotland means a permission granted by a road works authority to carry out road works. The equivalent definition applicable for England and Wales is given under 'street works licence'.
sewer authority:	in England and Wales, in relation to a public sewer, the sewerage undertaker within the meaning of the Water Act 1989 in whom the sewer is vested (section 89(1)(b)). In Scotland the equivalent provisions of the Act apply for 'a sewer vested in a local authority'.
street:	in England and Wales means the whole or any part of any of the following, irrespective of whether it is a thoroughfare -
	(a) any highway, road, lane, footway, alley or passage,
	(b) any square or court, and
	(c) any land laid out as a way whether it is for the time being formed as a way or not.

	Where a street passes over a bridge or through a tunnel, references to the street include that bridge or tunnel (section 48). The equivalent definition applicable in Scotland is given under 'road'.
street managers:	in England and Wales, means, in relation to a street which is not a maintainable highway, the authority, body or person liable to the public to maintain or repair the street, or if there is none, any authority, body or person having the management or control of the street (section 49). The equivalent definition applicable in Scotland is given under 'road managers'.
street with special engineering difficulties:	a street, or in Scotland a road, designated as such in accordance with section 63 or 122 of the Act.
street works:	works of any of the following kinds (other than works for road purposes) executed in a street, or in Scotland a road, in pursuance of a statutory right, a street works licence or a road works permission -

(a) placing apparatus, or

(b) inspecting, maintaining, adjusting, repairing, altering or renewing apparatus, changing the position of apparatus or removing it, or

(c) works required for or incidental to any such works, (including, in particular, breaking up or opening the street, or in Scotland the road, or any sewer, drain or tunnel under it, or tunnelling or boring under the street, or in Scotland the road) (sections 48 and 107).

street works licence:	in England and Wales, means a licence granted by a street authority to carry out street works. The equivalent definition applicable in Scotland is given under 'road works permission'.
traffic authority:	means in England and Wales the highway authority for the street concerned, and in Scotland the roads authority for the road concerned. (Based upon paragraph 70 of Schedule 8 to the Act.)
tramway:	a system, mainly or exclusively for the carriage of passengers, using vehicles guided, or powered by energy transmitted, by rails, or other fixed apparatus installed exclusively or mainly in a street, or in Scotland a road. (sections 105 and 164).
transport authority:	the authority, body or person having the control or management of a transport undertaking (sections 91 and 150).
transport undertaking:	a railway, tramway, dock, harbour, pier, canal or inland navigation undertaking of which the activities, or some of the activities are carried on under statutory authority (section 91 and 150).
undertaker:	the person in whom a statutory right to execute works is vested or the holder of a street works licence (see above) or the holder of a road works permission (see above), as the case may be.
verge:	means part of the highway, or in Scotland the public road, outside of the carriageway which may be slightly raised but is exclusive of embankment or cutting slopes, and generally grassed.